Social Attitudes in Northern Ireland

the 7th report

The Contributors

Paul Carmichael
Lecturer, Public Policy,
Economics and Law, University
of Ulster, Jordanstown.

Sally Cook
Research Officer, Environmental
Studies, University of Ulster,
Coleraine.

Ann Marie Gray
Lecturer, Social and Community
Sciences, University of Ulster,
Jordanstown.

Claire Guyer
Lecturer, Environmental Studies,
University of Ulster, Coleraine.

Deirdre Heenan
Lecturer, Social and Community
Sciences, University of Ulster,
Coleraine.

Joanne Hughes
Lecturer, Public Policy,
Economics and Law, University
of Ulster, Jordanstown.

Alan McClelland
Survey Manager, Central Survey
Unit, Northern Ireland Statistics
and Research Agency.

Martin Melaugh
CAIN Project Manager, INCORE,
University of Ulster and United
Nations University.

Adrian Moore
Lecturer, Environmental Studies,
University of Ulster,
Coleraine.

Niall Ó Dochartaigh
Lecturer, Political Science and
Sociology, University of Galway.

Gillian Robinson
Research Director, INCORE,
University of Ulster and United
Nations University.

Kate Thompson
Research Officer, Health
Sciences, University of Ulster,
Coleraine.

Social Attitudes in Northern Ireland

the
7th report

Editors
Gillian Robinson
Deirdre Heenan
Ann Marie Gray
Kate Thompson

Routledge
Taylor & Francis Group
LONDON AND NEW YORK

First published 1998 by Ashgate Publishing

Reissued 2018 by Routledge
2 Park Square, Milton Park, Abingdon, Oxon, OX14 4RN
711 Third Avenue, New York, NY I 0017, USA

Routledge is an imprint of the Taylor & Francis Group, an informa business

Notice:
Product or corporate names may be trademarks or registered trademarks, and are used only for identification and explanation without intent to infringe.

Publisher's Note
The publisher has gone to great lengths to ensure the quality of this reprint but points out that some imperfections in the original copies may be apparent.

Disclaimer
The publisher has made every effort to trace copyright holders and welcomes correspondence from those they have been unable to contact.

ISBN 13: 978-1-138-34501-0 (hbk)
ISBN 13: 978-1-138-34504-1 (pbk)
ISBN 13: 978-0-429-43813-4 (ebk)

Contents

List of Tables

List of Figures

Introduction

GILLIAN ROBINSON

This book is the seventh in a series of reports which began in 1991 (Stringer and Robinson, 1991; 1992; 1993: Breen, Devine and Robinson, 1995: Breen, Devine and Dowds, 1996: Dowds, Devine and Breen, 1997). As such it provides a unique opportunity to view social attitudes in Northern Ireland and explore how these may be developing and changing. In this volume authors from Northern Ireland and beyond utilise the data from the 1996 Northern Ireland Social Attitudes (NISA) survey to provide a picture of social attitudes to various issues. It is hoped that their analyses will stimulate others including policy-makers; journalists; community based groups; school children and students to further analyse the data, which is easily accessible (see Appendix 3), and gain a deeper understanding of social attitudes here.

The survey is constructed in a modular format and is closely linked to its sister survey British Social Attitudes (BSA) so that modules run in Northern Ireland are selected from those being run in GB. This allows for comparisons between Northern Ireland and GB and as you will see many of the authors chose to look at the differences and similarities between the two regions. In 1996 the issues covered included attitudes to the countryside and the environment; housing; and political trust. In addition there is a specific Northern Ireland module run each year and in 1996 this concentrated on aspects of community relations issues here. Finally the survey includes the International Social Survey Programme module that allows for comparisons with countries across four continents (over 20 countries in all) who participate in the programme. In 1996 this focused on the role of government. Full details on the content and administration of the survey are included in Appendix 1.

Having been involved with the Northern Ireland Social Attitudes survey series since its inception in 1989 I regret that this report will be the last in the series and presents the results of the final survey. The series began life with funding from the Nuffield Foundation and the Central Community Relations Unit (1989-1991) and since then funding has been provided by government departments in Northern Ireland. Unfortunately government is no longer willing to bear the full costs of the survey and efforts to secure other funding have been only partially successful. Therefore all those involved: Social and Community Planning Research (SCPR), who run the BSA series and were a key partner in the NISA series; government - CCRU and the Central Survey Unit within the Northern Ireland Statistics and Research Agency in particular; the University of Ulster (UU) team who have produced this report; and Lizanne Dowds, Research Affiliate at The Queen's University of Belfast have had to regretfully make the decision to end the series at this time.

However, as one door closes another one hopefully opens. Plans are underway to launch a new survey series which will retain some of the former NISA time series and on-going British Social Attitudes modules. Though the focus of the new survey will be quite different we are determined that the spirit of the original survey will not be lost and that the seven years data will continue to be built upon.

Acknowledgements

As the NISA survey series comes to a close the remaining task is to give a sincere thank you to all those who made the survey come to life and who sustained it throughout its lifetime. Peter Stringer was the academic who worked to get the survey extended to Northern Ireland and we thank him for his energy and interest. SCPR who enthusiastically agreed to extend the survey to Northern Ireland, and have directed the operation of the survey since then are we know saddened by its demise. We thank the BSA team for their support, interest and tireless efforts over the years. We are grateful to the Nuffield Foundation who provided funding over the first three years and to all the Northern Ireland government departments who have funded the survey over the years. In particular the CCRU has been an unflagging supporter of the series as have CSU who have carried out the fieldwork over the seven year period to their usual high standards. The research teams at The Queen's University of Belfast who produced the first six reports also deserve praise for their work to ensure the wider dissemination of the survey results. Many academics within Northern Ireland and beyond have contributed chapters to the series and a significant number of others have used NISA data in their research. Thank you all for your interest and we hope you will continue to find the data useful. The authors in this volume have worked to very tight deadlines and we thank you for your tolerance and co-operation. My colleagues at the University of Ulster who were prepared to take up the campaign to try to save the series and who are now working to secure funding for the new survey have been a tremendous support. In particular the analysis skills and thoroughness of Kate Thompson who prepared the data for authors was invaluable. Diane Devine prepared the many varied documents to camera ready copy and we are grateful for her dedication and unfailing good humour. Finally we would like to thank the people of Northern Ireland who participated in the survey each year and were so generous in their time and interest - without them no research of this kind is possible. We hope that the findings may in some small way contribute to the development and understanding of policy issues in Northern Ireland and may lead to policies that impact positively on the lives of all those living here.

References

Breen, R., Devine, P. and Dowds, L. (eds), (1996), *Social Attitudes in Northern Ireland: The Fifth Report,* Appletree Press, Belfast.

Breen, R., Devine, P. and Robinson, G. (eds), (1995), *Social Attitudes in Northern Ireland: The Fourth Report,* Appletree Press, Belfast.

Dowds, L., Devine, P. and Breen, R. (eds), (1997), *Social Attitudes in Northern Ireland: The Sixth Report,* Appletree Press, Belfast.

Stringer, P. and Robinson, G. (eds), (1991), *Social Attitudes in Northern Ireland: 1990-91 edition,* Blackstaff, Belfast.

Stringer, P. and Robinson, G. (eds), (1992), *Social Attitudes in Northern Ireland: The Second Report,* Blackstaff, Belfast.

Stringer, P. and Robinson, G. (eds), (1993), *Social Attitudes in Northern Ireland: The Third Report,* Blackstaff, Belfast.

1 Community Relations in Northern Ireland: Attitudes to Contact and Integration

JOANNE HUGHES AND PAUL CARMICHAEL

Introduction

The last ten years have seen a concerted effort on the part of successive UK governments to address the seemingly intractable problems of community relations in Northern Ireland. The approach has been twofold with attempts to find a constitutional resolution to the conflict complemented by the development of a community relations infrastructure. In the light of these developments, this chapter provides a résumé of the macro-political context over the period. It outlines the nature of community relations chiefly in terms of policy, legislative and infrastructural developments before offering an assessment of community relations in Northern Ireland. It goes on to consider evidence from the Northern Ireland Social Attitudes surveys conducted in 1989 and 1996. The central message is that the data indicate a discernible shift towards greater tolerance and mutual understanding.

Macro-political Developments

From 1987, a sustained twin-track approach to resolving Northern Ireland's constitutional imbroglio emerged. At the political level, the focus has been on attaining a constitutional settlement which will accommodate both unionists and nationalists. Complementing this, organisations with a dedicated community relations remit have been established and, through them, resources targeted at local community level. Arguably, it was the Enniskillen bomb on Remembrance Sunday in 1987 which prompted many in both communities to reconsider the future of Northern Ireland. In local government, for example, the incident was instrumental in promoting the development of 'responsibility sharing' in councils (Knox, 1996).   This move was consolidated by government measures to encourage inter-community cooperation at the local authority level.   Equally important were various European Community initiatives, notably, the European Community Peace and Reconciliation Programme (1995) which is designed explicitly to foster partnership and community-based forms of decision-making and service delivery.

Concurrent with these developments was the litany of the troubles. Though their intensity was at its lowest since during 1970-71, the relative decline in violence was shortlived. By 1989, an upsurge in loyalist violence triggered a familiar cycle of tit-for-tat reprisals by republican paramilitaries. By the early 1990s, these attacks were increasing in frequency culminating in the Provisional Irish Republican Army (PIRA) Shankill Road bombing of September 1993. The immediate outlook was bleak.

It was at this juncture that, in December 1993, the Downing Street Declaration between the British and Irish Governments marked a turning point. Whilst not the beginning of the end of the troubles, it appeared to be the end of the beginning, to coin a phrase. Hence, although the next eight months would witness further atrocities and outrages perpetrated by terrorists on both sides, by August 1994, the PIRA had declared a ceasefire, to be followed by the Combined Loyalist Military Command in October 1994. From thereafter, whilst the climate remained tense, the groundswell of goodwill and resolve to 'make peace work' was almost palpable. Notwithstanding genuine fears of a slide back into the worst excesses of the conflict, the period since 1994 was characterised by a widespread hope and expectation of improvement. Outwardly at least, a relaxed atmosphere descended on Northern Ireland as the immediate threat of violence receded.

Perhaps inevitably, the honeymoon period could not last. The summer of 1995 witnessed the first of three successive summers of discontent centred on a controversial march by members of the Portadown Lodge of the Orange Order from Drumcree parish church.   However, 'Drumcree 1', as it was subsequently dubbed, did not of itself spell the end of the ceasefires. This came in February 1996, with the Irish Republican Army's (IRA) Canary Wharf bombing in London. Even then, however, Northern Ireland was spared the resumption of fullscale terrorism and, at the time of the 1996 survey on which this report was prepared, confidence about the prospects for a speedy restitution of the ceasefires remained buoyant.

Community Relations in Northern Ireland

The outbreak of violence in 1969 which followed Catholic demands for civil rights forced the British Government at Westminster to play a more active role in the local politics of Northern Ireland[1]. As an interim step, troops were sent in to quell riots which had erupted in interface areas. A series of more long term reforms followed. These were designed to address the inequities which were a consequence of unionist control in the regional government of Northern Ireland and to address nationalist concerns. Issues for particular consideration were the voting arrangements for local authorities; the procedures for addressing electoral boundaries and the allocation of housing - all of which had previously been used to strengthen the powerbase of unionists (Whyte, 1990). In addition, the government made a commitment to examine relationships between the two communities and underlying causes of violence.

The main infrastructural outcome was the establishment of the Community Relations Commission and a ministry to oversee its work. Modelled on similar lines to the UK Commission for Racial Equality which deals with race relations issues, membership was drawn equally from both communities. Primary functions of the commission included:

'the encouragement of bodies active in promoting improved community relations, advice to government, the provision of educational and other programmes, and the commissioning or carrying out of research on community relations themes'. (Gallagher, 1995, p.29)

The initiative was not without its detractors. Hayes (1972) argued that locating governmental responsibility within a single, small ministry marginalised the issue. Community relations policy should inform government decisions at all levels where policy decisions impact upon both communities (housing, education, industry, law enforcement etc.). The strategy of 'mainstreaming' advocated by Hayes, whilst finding little support at the time, has become a feature of public policy in Northern Ireland during the 1990s as evidenced by Policy Appraisal and Fair Treatment (PAFT) guidelines (this is discussed in more detail later in the chapter).

By the mid-1970s the official community relations infrastructure had collapsed. Various explanations as to the cause of its demise have been advanced. At an official level, it was stated that the Power-Sharing Executive of 1974 (a joint system of government between Protestants and Catholics)[2] obviated the need for a Community Relations Commission. Some commentators, however, have suggested that this may have been a convenient excuse. The real reason lay in the fact that politicians were becoming increasingly suspicious of the community development strategy promoted by the Commission. Their concern was that a strengthened voluntary sector could provide an alternative basis for community leadership (Gallagher, 1995). Whatever the cause, community relations became a dormant issue for over a decade.

The sea change in the latter half of the 1980s which saw the return of community relations as a priority issue for policy-makers was prompted by several factors. These included the electoral rise of Sinn Féin after the hunger-strikes by Republican prisoners; external pressure to tackle community relations exerted on the British Government through the McBride campaign[3]; and, the Anglo-Irish Agreement (1985) which introduced a new dimension, in the form of consultation with the Government of the Irish Republic, to policy decisions on Northern Ireland. Of greatest significance, however, was a paper prepared for the Standing Advisory Commission on Human Rights (SACHR) by Hugh Fraser and Mari Fitzduff. It examined the history of community relations in Northern Ireland and considered ways in which difficult issues could be resolved (Gallagher, 1995). Developments in several areas ensued, namely, policy, legislation and infrastructure.

1. Policy

In 1992 government established an explicit community relations policy which had three primary aims: (a) to increase contact between Protestants and Catholics; (b) to encourage greater mutual understanding and respect for diverse cultural traditions; and, (c) to ensure that everyone in Northern Ireland enjoys equality of opportunity and equity of treatment (Department of Finance and Personnel and HM Treasury, 1992, p.142). The PAFT guidelines relating to equality, equity and fair treatment were published in 1994. The preamble reads as follows:

'Equality and equity are central issues which must condition and influence policy, taking in all spheres and at all levels of Government activity, whether in regulatory and administrative functions or through the delivery of services to the public'. (Northern Ireland Office, 1994)

Areas identified as relevant for PAFT proofing include: religion; gender; political opinion; marital status; having or not having a dependant; ethnicity; disability; age; and sexual orientation. Policy proposals when forwarded for ministerial decision must indicate that a PAFT appraisal has been undertaken.

In addition, departments are required to monitor the impact of their policy on designated groups and to provide relevant training for public sector managers. Finally, all departments are required to submit an annual report to the Northern Ireland Civil Service Central Secretariat outlining action taken to implement the PAFT guidelines.

2. Legislation

There have been three key developments over the last ten years. These have been designed to address institutionalised segregation in education, discrimination in the workforce, and equality and equity concerns in the formulation of public policy.   Specifically, with respect to education, the Department of Education for Northern Ireland (DENI) made provision for integrated education under the Education Reform Order (1989).   In particular it became possible for existing segregated schools to opt for integrated status through a parental ballot. Recognising, however, that most schools would retain their segregated status, the Order also provided that two cross curricular themes would become mandatory in the teaching of most academic subjects, namely, 'Education for Mutual Understanding' (EMU) and 'Cultural Heritage'.   EMU aims, *inter alia*, to help children learn to 'respect themselves and others' and 'to know about and understand what is shared as well as what is different about their cultural traditions' (Northern Ireland Curriculum Council, 1990).   Although cross-community contact is not viewed as compulsory to the achievement of these goals, it remains an optional strategy which teachers are encouraged to use.   Measures to tackle discrimination in the workforce were threefold. In 1989, the Fair Employment (NI) Act was passed. This was followed in the same year by the establishment of a Fair Employment Commission and a Fair Employment Tribunal to deal with cases of alleged discrimination.   Employers with more than 10 employees are required to register with the FEC and to monitor the religious composition of their workforce. It is illegal to discriminate indirectly and limited affirmative action policies to address imbalance in the religious composition of the workforce are permitted. The third area was PAFT, as outlined earlier.

3. Infrastructure

A community relations infrastructure was re-established.   Central to this was the creation of the Central Community Relations Unit (CCRU) in 1987. Reporting directly to the head of the Northern Ireland Civil Service (NICS), CCRU was charged with formulating, reviewing and challenging policy throughout the government system with the aim of 'bringing the two sides of the community towards greater understanding' (CCRU, 1991). All government departments were required to critically assess their policies and procedures to ensure that community relations considerations informed the delivery of key services in areas such as health, education, housing and economic development. Other aspects of the CCRU remit require the Unit to develop new ideas which would improve community relations and to support on-going efforts aimed at prejudice reduction. Several initiatives followed which endeavoured to improve contact between the two communities. In September 1987, the Department of Education for Northern Ireland (DENI) released £250,000 for the establishment of a Cross Community Contact Scheme. Administered through DENI, the scheme targets young people under the age of 19 and provides new resources for those already working in the area of peace and reconciliation. Funding criteria stipulate that activities seeking assistance should:

'… Improve cross-community understanding, be in addition to existing activity, be purposeful, and wherever possible, result in on-going contacts between young people from two communities'. (Northern Ireland Information Service, 1987, p.3)

The scale of applications and the subsequent involvement of more than one third of the schools in Northern Ireland prompted further developments. In February 1989, £2m was made available by government for the advancement of community relations objectives. Of this, £250,000 was used to extend the Cross Community Contact Scheme. The remainder contributed towards the establishment of two bodies. The first was the Cultural Traditions Group, headed by the controller of the BBC and charged with designing programmes in the arts, media and museums which would encourage constructive discussion on cultural traditions issues in Northern Ireland. Second, the Northern Ireland Community Relations Council (CRC) was formed in 1990 as a semi-autonomous government-funded public agency. CRC gained charitable status and served as a resource centre and focal point for groups and individuals working to improve community relations. Bloomfield (1997, p.65) remarked that CRC was, with hindsight, the major development of the era in community relations in Northern Ireland forming 'a public location at which the profile of community relations work could be raised'. Together, both CRC and CCRU 'signalled a new phase in the development of the [cultural] approach' and 'in some ways, these bodies mirrored the previous incarnations of the ministry and the commission of 1969-74, but there were equally significant differences from those precedents' (Bloomfield, *ibid*). Third, in 1989, the government offered financial assistance to district councils in Northern Ireland for the establishment of a community relations programme. Funding was conditional on cross-party support. The aim of this initiative was to involve local councillors and to develop a local area response to the community relations problem.

An Assessment of Community Relations in Northern Ireland

Changes at the macro-level over the last decade have resulted in the proliferation of a broad range of community relations initiatives, programmes and organisations. In an effort to examine effectiveness and to determine good practice, the CCRU has commissioned a series of macro-evaluations (Social Information Systems, 1994; Knox, *et al*, 1994; Knox and Hughes, 1996). These relate specifically to projects delivered directly through the CCRU, CRC and district council programmes. In all cases, community relations work has, to some degree, been experimental. Unlike targeted initiatives (EMU, integrated schools and the work of the Cultural Traditions Group) the recipient constituencies and the nature of projects are not predetermined. The approach is bottom-up and largely reactive to perceived need at grass-roots community level. Evaluators are broadly in agreement on the range of approaches embraced by the CRC and district council initiatives. These can be classified under the following six headings which, whilst not mutually exclusive, give a flavour of the diverse nature of activity:

Cultural Traditions

Projects in this category are concerned with advancing mutual respect for diverse cultural traditions. Much of the work promotes community relations objectives through common interest in cultural heritage and local history. Whilst most projects are cross-community, some are located in single identity communities where the primary aim is confidence building. Examples of activity supported include: seminars, conferences and workshops organised by local history associations, cultural awareness courses organised by Irish language groups and the Orange Order.

Community Development/Community Relations

This approach favours community relations work which underplays religious and political divisions between Protestants and Catholics and encourages co-operation on the basis of social, economic and environmental issues. The rationale is that community relations issues will emerge and can be addressed during the course of meeting agreed objectives in less contentious areas. Examples of organisations supported include: Co-operation North (which fosters economic links between the north and south of Ireland) and local community associations.

Reconciliation

This is a dedicated community relations approach where the aim is very specifically to improve intergroup awareness and to foster respect. Examples include: Corrymeela (a residential centre which provides a neutral venue for cross-community holidays, seminars and conferences) and the Peace People which offers a structured programme of community relations activity tailored to the needs of participant groups.

Reactive

Initiatives in this category emerged in response to specific terrorist incidents and threats. The approach is publicity driven and examples include: Enniskillen Together (founded by like-minded Protestants and Catholics to express revulsion at the bomb which killed 11 people in the town on Remembrance Sunday 1987) and the Peace Train Organisation (aimed at highlighting terrorist disruption of the main train link between Belfast and Dublin).

High Profile Community Relations

Projects in this category involve large-scale events often organised to engender 'first time' contact between Protestants and Catholics. The nature of the encounter tends to be largely superficial with little or no interaction between participants. Examples range from festivals and musical productions to dog shows and exhibitions.

Education and Personal Development

Here the focus is on generating confidence at an individual level and exploring those issues, germane to the conflict, which have affected people's lives. Examples include a respite cottage provided by the Ulster Quakers Group for women whose spouses are in prison; access to education courses for women who may have been disadvantaged by the effects of conflict; and training for those who are keen to facilitate cross-community projects.

In almost all of the categories, the strategy has been to improve community relations through cross-community contact. For the most part, assessments by the evaluators have been positive. Measured against government policy objectives for effective community relations, they found that all categories generated contact and some had the potential to promote attitudinal change.

In summary, a commitment to improving community relations has been comprehensively embraced by government. It is manifest not only in policy objectives and legislation but in the targeting of dedicated community relations resources. The following section offers an analysis of survey evidence on social attitudes.

Social Attitudes - Survey Evidence

Overall, the results of the survey in 1996 compared with that in 1989 show an improvement in the professed perceptions of the population concerning community relations. Altogether, of the questions posed of respondents in 1996, eight could be directly compared with identical or virtually identical equivalents from 1989.

Relations between Protestants and Catholics in Northern Ireland

Respondents were asked about relations between Protestants and Catholics. Would you say that relations are better than they were five years ago, worse, or about the same now as then? In response, Catholics, Protestants and the two groups combined demonstrated a decisive shift in their attitudes. Whereas the balance of opinion in 1989 was more worse than better, the belief that relations had improved became clear in 1996. In 1989, only 21 per cent of respondents thought that relations had got better in the past five years, whilst 28 per cent thought they had got worse. Almost half (47 per cent) perceived them as about the same. By 1996, whilst 42 per cent thought relations were about the same, only 11 per cent thought they were worse, with almost half

(46 per cent) believing them to have improved. In all cases, both Catholics and Protestants shared in the overall sentiment that relations had improved.

Table 1.1   What about relations between Protestants and Catholics? Would you say that they are better than they were 5 years ago, worse or about the same now as then?

| | 1989 | | | 1996 | | |
| | Total | Cath. | Prot. | Total | Cath. | Prot. |
	%	%	%	%	%	%
Better	21	23	20	46	47	44
Worse	28	31	26	11	10	11
Same	47	44	50	42	41	43
Other	2	2	2	-	-	-
Don't know	2	2	2	2	1	2
No answer	*	*	1	*	*	*

Table 1.2   And what about relations in 5 years time? Do you think relations between Protestants and Catholics will be better than now, worse than now, or about the same as now?

| | 1989 | | | 1996 | | |
| | Total | Cath. | Prot. | Total | Cath. | Prot. |
	%	%	%	%	%	%
Better	25	30	22	43	48	39
Worse	16	16	16	8	4	10
Same	54	51	56	42	43	41
Other	*	*	1	2	1	3
Don't know	5	4	5	6	4	7
No answer	1	*	1	-	-	-

The spirit of optimism was echoed in the follow up question on how relations would fare over the next five years (Table 1.2). In 1989, over half (54 per cent) of respondents indicated that they saw matters remaining about the same. A quarter (25 per cent) saw them improving whilst 16 per cent expected a deterioration. By 1996, those expecting relations to be better had moved into a slight majority (43 per cent) over those expecting things to remain about the same. Those expecting relations to deteriorate fell to just 8 per cent. In all cases, Catholics tended to be marginally more optimistic than Protestants although both groups shared a strong tendency for optimism. Indeed, if nothing else, the surveys confirm an optimism amongst people in Northern Ireland who are otherwise accustomed to recurrent disappointment.

Prejudice Towards Religious Denominations in Northern Ireland

Respondents were asked about prejudice towards religious denominations within Northern Ireland. Specifically, they were asked to indicate the perceptions of levels of prejudice towards Catholics. Perhaps not surprisingly, there was a marked divergence between Catholic and Protestant respondents on this issue. Generally, Catholics perceived higher levels of prejudice than Protestants.

Table 1.3   **... thinking of Catholics - do you think there is a lot of prejudice against them in Northern Ireland nowadays, a little, or hardly any?**

	1989			1996		
	Total	Cath.	Prot.	Total	Cath.	Prot.
	%	%	%	%	%	%
A lot	31	38	27	25	29	21
A little	41	46	37	49	54	45
Hardly any	22	11	31	23	13	31
Don't know	6	6	6	3	4	2
No answer	1	-	1	*	-	1

For instance, in 1989, whilst 38 per cent of Catholics perceived a lot of prejudice against Catholics, only 27 per cent of Protestants expressed a similar view. Conversely, whilst 31 per cent of Protestants suggested that there was hardly any prejudice, only 11 per cent of Catholics agreed. Similar differentials applied in 1996 although significantly, the proportions of both groups believing there to be a lot of prejudice fell appreciably to 29 per cent for Catholics and 21 per cent for Protestants. Clearly, over the period, whilst the discrepancies in perceptions over prejudice were maintained, both were on a declining trajectory.

The Proximity of Religious Denominations - Neighbourhoods, Workplaces, and Schools

The survey probed sentiments concerning attitudes of respondents to the proximity of other religious groups in a variety of social contexts, namely, neighbourhoods, workplace, and schools. In all three areas, a preference for mixed religious arrangements held sway. Moreover, over the intervening period, these sentiments have become more pronounced.

Table 1.4   If you had a choice, would you prefer to live in a neighbourhood with people of only your own religion, or in a mixed-religion neighbourhood?

		1989			1996	
	Total	Cath.	Prot.	Total	Cath.	Prot.
	%	%	%	%	%	%
Only own	23	18	27	14	11	17
Mixed	70	75	67	82	85	80
Don't know	5	6	5	4	5	3
No answer	1	1	2	-	-	-

Respondents were asked to say whether they would prefer to live in a neighbourhood which was exclusively populated with people of their own religious persuasion or in an area with residents of mixed religions. In 1989, fully 70 per cent of respondents indicated a preference for mixed religious neighbourhoods although almost a quarter (23 per cent) preferred neighbours of their own faith only. By 1996, however, this had fallen to just one in seven (14 per cent) of all respondents (but 17 per cent of Protestants). On the other hand 82 per cent now expressed a preference for mixed neighbourhoods.

A note of caution might be injected here when interpreting these figures. For example, consider the relative proportions of each religious group in the overall composition of an area - the survey does not indicate respondents' thoughts with this degree of detail even though it is probably crucial. For instance, a ratio of 20:80 in the religious composition of a defined area is mixed but so preponderant is the majority group that the continued willingness of members of the minority group to remain *in situ* is heavily compromised. This is reflected in the increasingly spatial polarisation of the population of Northern Ireland since the resumption of the troubles as, for example, in the 'retreat' of Protestants from many border areas, and to the eastern parts of

Northern Ireland generally. In a very real sense, therefore, the professed willingness (even alacrity) of respondents for mixed living contrast vividly with the reality of trends on the ground.

Table 1.5   And if you were working and had to change your job, would you prefer a workplace with people of only your own religion, or a mixed-religion workplace?

	1989			1996		
	Total	Cath.	Prot.	Total	Cath.	Prot.
	%	%	%	%	%	%
Only own	11	7	14	3	2	4
Mixed	83	86	81	96	97	95
Don't know	5	6	3	2	2	1
No answer	1	1	2	-	-	-

Respondents were asked to indicate their views on a mixed or exclusive religious composition in their workplace (Table 1.5). The pattern was ostensibly one of an increasing desire for religious mix. Indeed, the pattern was even more pronounced than with neighbourhoods. Both groups shared a strong preference for workplaces with a mixed religious composition. Again, we might speculate about what is implied here. Essentially, most people at work find themselves in environments in which they will, almost unavoidably, be in invariably regular and often close contact with people from various religious backgrounds. Indeed, the crucial role of work in the diurnal pattern of most of the adult population makes contact inevitable. Moreover, the impact of the legislation on the display of flags and emblems (the Public Order ((Northern Ireland)) Order 1987) makes workplaces less intimidating for religious minorities.

Of course, whilst contact in the working environment can help to demolish barriers and hostility between the two communities, it is, perhaps, in the early years, that these obstacles to social stability and inter-group harmony are inculcated. To that extent, the existence of an education system overwhelmingly characterised by its fundamental bifurcation based on religion has long been held to help perpetuate division (Whyte, 1990).

Table 1.6 shows significantly that just over half of respondents (53 per cent) in 1989 supported co-religious schooling, whilst two fifths (39 per cent) preferred to send their children to schools with children of only their own religion.

Table 1.6   **And if you were deciding where to send your children to school, would you prefer a school with children of only your own religion, or a mixed-religion school?**

	1989			1996		
	Total	Cath.	Prot.	Total	Cath.	Prot.
	%	%	%	%	%	%
Only own	39	37	41	34	38	31
Mixed	53	54	52	62	57	65
Don't know	8	9	6	5	6	4
No answer	1	-	1	-	-	-

Indeed, more Protestants (41 per cent) than Catholics (37 per cent) were in favour of segregated schooling. While the Catholic Church has a long-established school sector (the 'Maintained' sector where almost all pupils are Catholic), there are very few/no equivalent Protestant schools with, instead, a 'Controlled' sector of schools with an overwhelming preponderance of Protestant pupils. This is not a matter of semantics. Indeed, it is ironic considering that the debate over segregated schooling was, for a long time, conducted around the assertion that such segregation existed and persisted at the sole behest of the Catholic Church. By 1996, the preference of Catholics for segregated education had remained almost static (38 per cent), whilst that of Protestants had fallen to 31 per cent. Mixed schooling was the preferred choice of 62 per cent of the overall population. Hence, whilst the integrated school sector in Northern Ireland is very small, the sector has seen considerable growth, now accounting for some 2 per cent of pupils. An important development in this regard has been the improved financial arrangements for integrated schools following the government's decision to allow 100 per cent funding after twelve months if the school seeking funding can demonstrate its viability (Donnelly, 1997).

Employment Chances

On the subject of employment, respondents were asked to indicate whether they thought that the chances of either a Protestant or a Catholic getting a job were the same or different. Significantly, in 1989, in the most polarised set of responses, a large majority of Catholics believed that the chances were different (that is, Catholics were less likely to get a job) whilst an identical proportion of Protestants felt that the chances were the same.

Table 1.7   Thinking now about employment. On the whole, do you think the Protestants and Catholics in Northern Ireland who apply for the same jobs have the same chance of getting a job or are their chances of getting a job different?

	1989 Total	1989 Cath.	1989 Prot.	1996 Total	1996 Cath.	1996 Prot.
	%	%	%	%	%	%
Same	47	30	60	50	38	59
Different	43	60	30	43	57	33
Don't know	10	11	10	7	5	8
No answer	1	-	1	*	-	*

Of all the areas canvassed, it is well known that attitudes about discrimination in employment whether based on truth or perception are particularly emotive. By 1996, the picture remained sharply polarised albeit with a decline in the proportion of Catholics perceiving different chances along with a considerable rise (to 38 per cent) in the proportions thinking that the chances were the same. Of interest, however, was the changing reaction of Protestants over the period, with a slight but perceptible rise in the numbers believing chances to be different (up to 33 per cent).

Speculation as to why this is so is just that - speculation. However, it could be connected with the growing evidence of Protestant alienation. *Inter alia*, this is a belief amongst Protestants that it is they who are now the disadvantaged community in the labour market thanks to fair employment legislation and the general perception of a Catholic population in its stride which, in the zero-sum game of Northern Ireland politics, translates directly into an 'obvious' retreat for Protestants. Evidence for such conjecture may be gleaned from considering the evidence of a further survey question which asked respondents to indicate which group is more likely to get a job (Table 1.8). Again, the results are revealing. Whilst they indicate that half (50 per cent) of the respondents think the chances are the same for both communities, the proportion thinking that it is Protestants who are advantaged fell from 32 per cent in 1989 to 26 per cent in 1996. Moreover, the decline in such sentiments was more pronounced amongst Catholics, suggesting a growing faith in the ability of the legislation and changed attitudes to deliver greater equality of treatment. In much the same way, the proportion believing it to be Catholics who are the advantaged group rose (from 9 per cent to 12 per cent), a belief shared by a sharply higher number of Catholics in 1996 compared with 1989 (6 per cent, up from less than 1 per cent).

Table 1.8   Which group is more likely to get a job - Protestants or Catholics?

| | 1989 | | | 1996 | | |
	Total %	Cath. %	Prot. %	Total %	Cath. %	Prot. %
Same chance	47	30	60	50	38	59
Protestants	32	59	12	26	46	10
Catholics	9	*	16	12	6	17
Depends	4	3	5	-	-	-
Don't know	4	4	4	12	11	13
No answer	4	4	4	*	-	*

Conclusion

When asked the question 'Have community relations improved?', it is difficult to provide a definitive 'yes' or 'no'. Whilst Fitzduff (1993) offered a synopsis of measures taken to improve community relations, Bloomfield concludes that Fitzduff's synopsis is indicative of 'trends rather than irrevocable shifts' which are 'difficult to prove or disprove'. Bloomfield added that, whilst 'most people in Northern Ireland would probably endorse the general trend of Fitzduff's comments', it is 'more difficult to assess the question of any causal link between the CRC's establishment and the noted improvements' (Bloomfield, 1997, pp.144-45).   Nonetheless, ostensibly, evidence from comparative analysis of two surveys in Northern Ireland on attitudes to contact and integration offers cautious grounds for optimism. Some initiatives have had a direct impact on attitudes, for example, legislation of fair employment and flags and emblems.   Equally, albeit slowly, there is now a momentum behind the concept of integrated education. With the erosion of the power and influence of the churches, coupled with the growing reputation for standards of the integrated schools, not least as serious alternatives for children failing to enter grammar schools, integrated schooling is now an established feature in Northern Ireland.   All of these developments have been underpinned by extending responsibility sharing in government, the support forthcoming from various European Union sponsored initiatives, as well as progress towards securing a wider constitutional settlement for Northern Ireland.

Survey evidence illustrates that public *attitudes* on a range of issues associated with improving community relations, moved decisively towards closer inter-communal association and integration. Paradoxically, however, what this evidence does not reveal is the discernible shift in the electoral *behaviour* of people in Northern Ireland, with increasing polarisation becoming apparent (Knox and Carmichael, *forthcoming*). Arguably, Northern Ireland is becoming more, rather than less, polarised. As the summers of 1995, 1996 and 1997 attest, beneath the veneer of improving relations and emerging harmony, many of the old prejudices, suspicions and hatreds remain. Atavistic tendencies within both traditions resurfaced with alarming speed in the most graphic form, with one of the worst bouts of civil unrest in the troubled history of the Province - Drumcree. Harangued from all sides for having failed, the whole episode left in turmoil those engaged in community relations, forcing many to go back and re-examine first principles. Hence, whether the improvement in attitudes and *some* of the behavioural patterns discerned in our survey analysis above will be replicated on a long term sustainable basis throughout Northern Ireland society remains to be seen.

Notes

1. Prior to 1969, a convention had been established in Westminster such that all Northern Ireland business, except for a few limited areas of policy, was within the remit of the region's parliament in Stormont (Belfast). By 1972, it was clear that the devolved Northern Ireland Government was unable to control the civil unrest. Invoking its powers under the Government of Ireland Act, the Westminster parliament suspended the Northern Ireland Government and replaced it with direct rule from Westminster.
2. The power-sharing executive lasted three months and remains Northern Ireland's only experience of internal government based on a sharing of power between the two communities. Its supporters aimed to construct a devolved system based on power-sharing and a joint body, the Council of Ireland, to regulate a small number of largely minor affairs of common concern to the north and south of Ireland. It was opposed by the Democratic Unionist Party and most of the Ulster Unionist Party. Eventually it was brought down by the Protestant workers' strike of 1974.
3. The McBride campaign, initiated by Irish Americans, sought to prevent US owned companies from investing in Northern Ireland on the grounds that the British Government was not doing enough to eliminate discrimination against the Catholic population.

References

Bloomfield, D. (1997), *Peacemaking Strategies in Northern Ireland*, Macmillan, Basingstoke.

Central Community Relations Unit. (1991), *Community Relations in Northern Ireland*, CCRU, Belfast.

Department of Finance and Personnel and H.M Treasury. (1992), *Northern Ireland Expenditure Plans and Priorities 1992-93 to 1994-95*, DFP, Belfast.

Donnelly, C. (1997), *Ethos and Governance : Case Studies from the Primary School Sector in Northern Ireland*, Unpublished PhD Thesis, University of Ulster, Jordanstown.

Fitzduff, M. (1993), *Uneasy Partners? Government, Community Relations, and the Voluntary Sector*, Community Relations Council, Belfast.

Gallagher, A.M. (1995), 'The Approach of Government: Community Relations and Equity', in S. Dunn, (ed), *Facets of the Conflict in Northern Ireland*, St Martin's Press, New York.

Hayes, M. (1972), *The Role of the Community Relations Commission in Northern Ireland*, Runnymede Trust, London.

Knox, C. (1996), 'The Emergence of Power Sharing in Northern Ireland : Lessons from Local Government', *Journal of Conflict Studies*, vol.16, no.1, pp.7-29.

Knox, C. and Carmichael, P. (forthcoming), 'The Northern Ireland Local Government Elections 1997', *Government and Opposition*.

Knox, C., Hughes J., Birrell, D. and McCready, S. (1994), *Community Relations and Local Government*, Centre for the Study of Conflict, University of Ulster, Coleraine.

Knox, C. and Hughes, J. (1996), *Community Relations: A Research Review*, Centre for Research in Public Policy and Management, University of Ulster, Jordanstown.

Northern Ireland Curriculum Council. (1990), *Cross-curricular Themes - Guidance Materials*, NICC, Belfast.

Northern Ireland Information Service. (1987), 'Community Relations', *Minister's Press Statement*, 14 September.

Northern Ireland Office. (1994), *Policy Appraisal and Fair Treatment Guidelines*, Mimeo.

Social Information Systems. (1994), *A Review of Community Relations Research in Northern Ireland*, SIS Ltd, Cheshire.

Whyte, J. (1990), *Interpreting Northern Ireland*, Clarendon Press, Oxford.

2 The Growth of Home Ownership: Explanations and Implications

DEIRDRE HEENAN

The 18-year period of Conservative Government from 1979 to 1997 witnessed profound changes in British housing policy. It could be said that the Thatcherite principles of competition, promotion of the free market, and privatisation can be more clearly identified in housing policy than any other area of social policy. Privatisation was introduced to housing on a scale never before imagined. Local authorities moved from their traditional role of direct providers of housing to enablers, enabling a range of other organisations such as housing associations to provide housing services. The cornerstone of this privatisation programme has been the vigorous promotion of home ownership. The Northern Ireland Social Attitudes survey (NISA) series regularly includes a section on housing issues and in 1996 the focus was on home ownership. Respondents were asked their views and opinions on this form of tenure. This chapter assesses the responses to these questions and places them in the context of overall changes in housing policy.

Tenure

Owner occupation is overwhelmingly the preferred tenure in Northern Ireland. Three-quarters of NISA respondents were owner occupiers. This figure is slightly higher than the comparable figures from other Northern Ireland surveys such as the 1997 Family Expenditure Survey and the 1997 Northern Ireland Housing Executive (NIHE) House Condition Survey.

Table 2.1   Comparison of data on tenure from three Northern Ireland surveys

	owner occupation %	public renting %
1996 Social Attitudes Survey	75	19
1997 Family Expenditure Survey	66	20
1997 House Condition Survey	66	23

The high levels of owner occupation can be seen as a direct result of government housing policy which has vigorously promoted privatisation. This privatisation has taken a number of forms but in terms of numbers the most important measure was the introduction of.the 'right to buy', i.e. the sale of council houses to individual tenants. The Housing Act (1980) in England and Wales and the Housing (Tenants' Rights etc.) Scotland Act (1980) gave local authority tenants the legal right to buy their own homes at a substantial discount on the market value. Initially discount was up to a maximum of 50 per cent, though this was extended through subsequent Acts to 60 per cent on houses and 70 per cent on flats.

In Northern Ireland housing is administered by the NIHE which was established in 1971 when housing functions were removed from local government following allegations of political and religious discrimination, and placed with this new government quango. The NIHE was established as a single-purpose, efficient and streamlined central housing authority, to take over responsibility for the building, management and allocation of all public sector housing.

The NIHE was promoting house sales before the government introduced the right to buy legislation in the UK. In May 1979 it introduced a voluntary house sales scheme. Under this scheme a secure tenant of at least three years, wishing to buy, was entitled to a discount of up to 50 per cent. A number of dwellings such as bungalows with one or two bedrooms and sheltered properties were excluded from the scheme. This scheme operated until September 1983, when the Housing (Northern Ireland) Order introduced the right to buy. The NIHE amended the voluntary sales scheme to take account of the additional rights contained within the Order. Over 60,000 sales have been completed through the NIHE house sales scheme. Since the early 1990s however figures for sales have levelled off and owner occupation is currently increasing at approximately 1 per cent per annum at the expense of the rented sectors (NIHE, 1996).

Table 2.2   Increase in owner occupation in Northern Ireland

	owner occupied	rented, NIHE
	%	%
1981	54.04	37.89
1984	57.81	35.74
1986	60.70	33.76
1990	65.29	29.18
1993	67.77	26.72

Source: Dept. of the Environment (NI), *Housing Statistics 1993*.

Since the late 1980s the NIHE has moved away from being a direct provider of new housing to an enabling role, assisting the private and voluntary sectors. Provision of new build by the NIHE has diminished from 71 per cent in 1976 to 13 per cent in 1993 (DOE, 1993). This substantial decrease in house building activity has resulted in it moving from being by far the largest provider of new housing, to being on a par with housing associations (Melaugh, 1994). Whilst in GB local authorities operating right to buy schemes were forced to forward 80 per cent of capital receipts from sales back to central government, the situation in Northern Ireland has been somewhat different. The NIHE has been allowed to retain capital receipts from sales to reinvest in the housing programme. It has been suggested that this unique financial arrangement has resulted in the NIHE adopting a relatively rigorous approach to the promotion of house sales (Singleton, 1982).

Sales to Tenants

Eighteen per cent of owner occupiers included in NISA had bought their own homes as NIHE tenants. This is broadly comparable with the figure produced by the NIHE. It is however important to note that purchasers of NIHE properties are not typical of NIHE tenants as a whole. Those who took advantage of the right to buy their homes were usually middle aged, skilled workers with a grown up family. Generally there were at least two earners in the family and no pensioners. They were more likely to be in full time employment and on higher incomes than a non-purchaser (NIHE, 1986). This is similar to the profile of a typical purchaser in GB (Forrest and Murie, 1988). Also the dwellings sold were most likely to be semi-detached and built in the post-war period. The best houses in the best locations were sold first leaving an increasing proportion of less popular estates in the public sector (NIHE, 1986).

Those who purchased their homes could be described as the top layer of NIHE tenants, those in the most favourable economic position. The removal of these individuals has resulted in a narrowing of the remaining tenant base. It is less mixed in terms of age, social class, marital status and employment status. Tenants are more likely to be dependent on state benefits, i.e. the unemployed, the elderly and lone parents.

Table 2.3   Average gross weekly household income by tenure of dwelling

		NI		UK
	1993/94	1994/95	1995/96	1995/96
	£	£	£	£
NIHE/local authority	143.91	166.06	179.89	179.15
Owned with mortgage	468.12	463.00	484.09	542.50
Owned outright	268.23	359.07	287.00	327.96
All dwellings	299.23	326.32	322.76	380.89
Base number of households	655	628	659	6,797

Source: Family Expenditure Survey 1997

In both Northern Ireland and GB income is highest for households living in accommodation owned with a mortgage and lowest for households in the public rented sector. In 1995/96 the average gross weekly income of Northern Ireland households whose accommodation was owned with a mortgage was almost three times as great as the income of those living in the public rented sector.

A further measure to promote owner occupation, particularly low cost home ownership, was the establishment of the Northern Ireland Co-ownership Housing Association (NICHA). NICHA was set up in 1978 to promote and develop the concept of equity sharing in the private housing sector. The co-ownership scheme enables prospective purchasers to enter owner-occupation, where they would otherwise be unable to afford it. Eventually when household income increases shared owners can move into full ownership by buying the remainder of the property or by selling their share and moving on to another property. Since its formation it has enabled 13,000 households to move into home ownership, of these 7,800 have progressed to 100 per cent equity share (DOE, 1996).

Rent increases also appear to be an important factor in encouraging house purchase. During the first Thatcher administration, council rents rose by 119 per cent, during the same period the retail price index rose by 55 per cent (Malpass, 1992). The Conservative Government stated that council rents should equate to private sector rents and subsequently rents in Northern Ireland have risen faster than inflation (NIEC, 1996). In reality, then, it is often cheaper to buy your home than to rent it. Unsurprisingly then, almost 90 per cent of respondents either agreed or strongly agreed that owning your own home was cheaper than renting. This can be compared to an NIHE survey in 1996 which reported that of those who had applied to buy their NIHE properties, 58 per cent claimed recent rent rises and the fact that mortgage payments were cheaper than renting were very important reasons for purchase.

Table 2.4   Over time, buying a home works out less expensive than paying rent

	%
Agree strongly	55
Just agree	35
Neither agree/disagree	5
Just disagree	3
Disagree strongly	1

Alongside these rent rises the housing benefit system was reformed. In 1983 households on 110 per cent of average male earnings had been eligible for housing benefit, but by April 1988, only households on less than 50 per cent of average male gross earnings were eligible to claim. The substantial reduction of eligibility was another powerful incentive for tenants not eligible for Housing Benefit but facing rent increases to move to owner occupation, either through the right to buy or the private market (Ungerson, 1994). Additionally it has led to those in receipt of housing benefit being concentrated in the public sector. In Northern Ireland 76 per cent of all NIHE tenants are receiving housing benefit and the vast majority, 92 per cent, of new NIHE and Housing Association tenants are in receipt of Housing Benefit (DOE, 1996).

Table 2.5   Regional average house prices, mortgage advances and income of borrowers for 1996

Region	average dwelling price £	average advance £	average recorded income of borrowers £
North	51,684	39,135	19,530
Yorkshire and Humber	55,867	42,733	21,058
East Midlands	58,855	43,462	21,749
East Anglia	61,819	45,709	22,493
Greater London	94,065	68,908	31,482
South East (exc. Greater London	85,767	60,745	28,629
South West	68,034	48,218	22,935
West Midlands	64,320	46,693	23,077
North West	57,701	44,114	21,365
Wales	54,898	41,584	20,927
Scotland	56,674	42,775	21,676
Northern Ireland	47,678	35,830	19,197

Source: Survey of Mortgage Lenders 1997

An additional attraction of home ownership in Northern Ireland is the fact that house prices here are the lowest in the UK and the market has not witnessed the fluctuations experienced in the rest of the UK. House prices have remained relatively stable, rising steadily since 1980 and unlike GB they have increased at a slower rate than earnings. Consequently home owners have been protected from problems of negative equity and relatively smaller mortgages have meant that repossessions have not become a feature of the housing market. The stability of the market has meant that houses are relatively more affordable in Northern Ireland. In 1994 the Affordability Ratio produced by the Council of Mortgage Lenders was the best in the Province for over ten years (DOE, 1996).

Rent or Buy?

Respondents were asked whether, if they had a free choice, they would rent or buy their home. The vast majority, 88 per cent, of respondents reported that they would buy their home.   This must in part relate to the recent residualization of public housing.   The marked social and spatial divisions between tenants and home owners means that the status of council housing has been lowered and the stigma attached has increased (Malpass and Murie, 1996).   The unpopularity of the private rented sector could in part be related to the fact that because it accounts for such a small percentage of the stock in Northern Ireland it is not considered to be a realistic alternative. The private rented sector currently accounts for just 3 per cent of the stock compared to 10 per cent in GB.   In Northern Ireland home ownership is considered to be the ideal with renting either from the private sector or from the NIHE being seen as a poor second.   As Ungerson (1994, p. 209) notes 'one of the great successes of Conservative housing strategy in the 1980s was to bring the discourse of ownership into absolute predominance'.

Table 2.6   If you had free choice would you rent or buy?

	%
Would choose to rent	11
Would choose to buy	88

Reasons for Not Purchasing

Those who had not entered into home ownership were asked to give reasons for not purchasing their own homes.   The most popular reason for not buying one's own home was that respondents had not got a secure job.   This was closely followed by not being able to afford the property they would like.   There was a clear relationship between those who were unable to buy the property for financial reasons and those who were currently NIHE tenants.   Of those who could not afford the property 84 per cent were NIHE tenants, whilst 88 per cent of those who could not afford the repayments were NIHE tenants.

This again confirms the earlier profile of NIHE tenants.   They are more likely to consider home ownership to be out of their reach.   The disadvantage suffered by those citizens unable to enter occupation because of their employment status is summed up by Murie (1993, p. 143).

'Access to home ownership generally involves negotiation of a loan. Eligibility for a loan relates to employment and occupation and the size of loan to income. Those outside the labour market or on low wages will not be in a position to buy. Affordability problems are not a new phenomenon, but as home ownership includes a larger part of the housing system ineligibility for that tenure is more limiting.'

Table 2.7   Reasons for not purchasing *

	%
Do not have a secure enough job	71
Cannot afford property would like	70
Not able to afford deposit	67
Difficult to keep up repayments	66
Not able to get mortgage	63
Might be unable to resell when wanted	35

*   could give more than one reason

Advantages of Home Ownership

Respondents were asked the main advantages of home ownership. As already discussed, a powerful incentive for buying one's own home is the fact that buying is cheaper than renting as house prices are relatively low. The second most popular advantage of home ownership is having something to leave to one's family. The inheritance value of home ownership is something which has been highlighted in similar studies in GB. In his study of home ownership in the UK Lynn (1991) noted similar results. The most popular reason for buying one's own home, cited by 76 per cent of his respondents, was because buying was considered to be a good financial investment.

Table 2.8   Advantages of home ownership

	agree strongly %	agree %
Over time, buying a home works out less expensive than paying rent	54	35
Owning your home makes it easier to move	22	32
Owning your home gives freedom to alter it	35	37
Can leave home to the family	50	36

The next most popular reason for home ownership reported in his study was having something to leave to the family, which was forwarded by 56 per cent of respondents. The importance of the inheritance value of owner occupation was highlighted in the 1992 Conservative Party Manifesto (p. 33) which stated,

'The opportunity to own a home and pass it on is one of the most important rights an individual has in a free society. Conservatives have extended that right. It lies at the heart of our philosophy'.

While the majority of respondents appeared content with home ownership a small number did highlight a range of disadvantages. By far the most cited disadvantage of home ownership given was the financial cost of maintaining a home.

Table 2.9   Disadvantages of home ownership

	agree strongly %	agree %
Owning your home can be a risky investment	8	27
Owning a home ties up money you may need urgently for other things	6	24
Owning a home is a big financial burden to repair and maintain	13	35
Owning a home is just too much of a responsibility	3	8

This may in part be related to the fact that there is now a significant number of home owners who can just barely afford to be in that tenure group. As the levels of home ownership have increased, the home owning population has become much more diverse. Those owning their own homes include those who own relatively cheap properties as well as affluent home owners. For many households owner occupation is the only option, they have little alternative but to enter at the lower end of the market. The shrinking public sector means they have little chance of being allocated public housing. The size of the private rented sector limits its role as a realistic alternative. Clearly, therefore, it is difficult to talk in any coherent way about home owners as a homogeneous group and the label 'home owner' is no longer synonymous with wealth. An increasing proportion of home owners are there through lack of choice and are finding the upkeep and maintenance of their homes a financial strain.

Views on their Area of Residence

A series of questions relating to tenants attitudes to the area they lived in were included in the study. Respondents were asked whether their area had got better, worse or remained the same in the past two years (Table 2.10). The majority of respondents (58 per cent) reported that their area had stayed the same, 23 per cent thought it had improved, while 18 per cent thought it had got worse. Sixty-three per cent thought it would stay the same over the next two years, 20 per cent thought that it would get better, while 15 per cent thought that it would get worse. Levels of satisfaction with one's area of residence appear to have increased during the decade as in 1990, just 17 per cent of respondents thought their area had improved and only 15 per cent were positive about the future.

Attitudes towards area of residence were clearly related to the tenure status of the respondents. Owner occupiers were much more likely to have a positive view of their area than those who rented their homes. Some 23 per cent of NIHE tenants thought that job opportunities in their area were worse than average compared to 7 per cent of owner occupiers. When asked about schools in their area 37 per cent of owner occupiers thought the schools in their area were better than average compared with 32 per cent of NIHE respondents.

Table 2.10   **Respondents' attitudes to whether their area of residence had got better, worse or remained the same during the last two years**

	1990 %	1996 %
Better	17	23
Worse	12	18
About same	71	58

Given the spatial disadvantage suffered by many NIHE tenants these results are hardly surprising. In GB Massey (1994) noted that newer industries have been located in regions and cities where council housing is in short supply. Hallsworth, *et al* (1986) commented on the fact that many local authority housing schemes contain few if any basic shopping facilities and public transport is limited. Finally tenants suffer from a more general tendency for children from poorer areas to attend schools with below average facilities and less well qualified and motivated teachers (Kirby, 1979).

Table 2.11   **Respondents' priorities for extra government spending**

	%
Health	63
Education	21
Social security benefits	4
Housing	4
Help for industry	3
Roads	2
Public transport	1
Police and prisons	1
Other	1

It is interesting to note that the cuts in public expenditure on housing have not impinged on the public consciousness to the same extent that health and education cuts have. Respondents were asked to identify their first priority for extra government spending. Housing was identified as the most important priority by only 4 per cent of respondents. Despite the fact that housing was the particular area singled out for government cuts, health was overwhelmingly the most popular choice for extra spending, suggested by 63 per cent of the respondents.

Conclusion

In conclusion then, while owner occupation is the preferred tenure in Northern Ireland, much of the increase in this tenure since the early 1980s can be seen as a direct result of government policy. The promotion of owner occupation has been unrelenting and unprecedented. In effect government has been subsidising the better off to buy their own homes at the expense of the less well off. The gains in the housing market have not been evenly distributed and the transfer of housing wealth has led to deeper divisions between those who own their homes and those who do not (Munroe, 1988). Government housing policies have rewarded those on higher incomes and helped home owners accumulate wealth in a way in which tenants are excluded. This has resulted in increasing polarisation, local authority housing has become residualised and is widely viewed as suitable only for those who are unable to enter the private market. Through Conservative Government housing policies society is becoming more unequal. The poorest families are being pushed onto estates where they are allocated the poorest housing. They are trapped in a cycle of deprivation, the main features of which are: poor housing; poor environment; poor schools and poor job prospects (Dean, 1997). The reduced social mix of the tenants has lessened their bargaining power and therefore their ability to do something about their circumstances.

The figures on the growth of home ownership appear to suggest that sales have reached a peak and there is a ceiling on owner occupation. The high rate of sales to sitting tenants achieved at the inception of the right to buy policy have not been maintained. The very high levels of NIHE and housing association tenants in receipt of Housing Benefit suggests that the proportion who have the ability to buy is decreasing. Therefore the alternatives to owner occupation must be maintained and invested in by government.

Compared to other areas of social policy, such as health and crime, housing has lagged behind on the political agenda. Yet it is clearly a central component in the success of these other areas. A long term, coherent housing policy which recognises the value of all tenures is something which is urgently required. A housing policy which is based on privatisation at the expense of all other options has had its day, what is needed is a new vision.

References

Council of Mortgage Lenders and the Department of Environment Transport and the Regions. (1997), *Survey of Mortgage Lenders*, London.

Dean, M. (1997), 'Tipping the Balance', *Search*, Issue 27, Spring.

Department of the Environment. (1993), *Housing Statistics*, DOE, Belfast.

Department of the Environment. (1996), *Building on Success: Proposals for Future Housing Policy*, DOE, Belfast.

Forrest, R. and Murie, A. (1988), *Selling the Welfare State: the Privatisation of Public Housing*, Routledge, London.

Hallsworth, A.G., Wood, A. and Lewington, T. (1986), 'Welfare and Retail Accessibility', *Area*, 18, 4, pp. 291-298.

Kirby, A. (1979), *Education, Health and Housing*, Saxon House, Farnborough.

Malpass, P. (1992), 'Housing Policy and the Disabling of Local Authorities', in J. Birchall, (ed), *Housing Policy in the 1990's*, Routledge, London.

Malpass, P. and Murie, A. (1996), *Housing Policy and Practice*, (4th ed), Macmillan, London.

Massey, D. (1984), *Spatial Divisions of Labour*, Macmillan, London.

Melaugh, M. (1994), 'Housing and Religion in Northern Ireland', *Majority, Minority Review No.3*, Centre for Study of Conflict, University of Ulster, Coleraine.

Munroe, M. (1988), Housing Inheritance and Wealth, *Journal of Social Policy*, vol. 17, no.4, pp. 417-436.

Murie, A. (1993), 'Restructuring Housing Markets and Housing Access', in R. Page and J. Baldock, (eds), *Social Policy Review 5: The Evolving State of Welfare*, Social Policy Association, Canterbury.

Northern Ireland Economic Council (1996), *Building a Better Future*, NIEC, Belfast.

Northern Ireland Housing Executive. (1996), *Housing Executive Voluntary Sales Survey 1992-1995*, NIHE, Belfast.

Northern Ireland Housing Executive. (1986), *House Sales Review and Potential Future Demand*, Housing and Planning Research Unit, June.

Lynn, P. (1991), *The Right to Buy: A National Follow-up Survey of Council Homes in England*, HMSO, London.

Policy Planning and Research Unit (1997), *Family Expenditure Survey*, PPRU, Belfast.

Singleton, D. (1982), 'Council House Sales in Northern Ireland', *Housing Review*, March-April, pp. 43-45.

Ungerson, C. (1994), 'Housing: Need, Equity Ownership and the Economy', in V. George, and S. Millar, (eds), *Social Policy Towards 2000; Squaring the Welfare Circle*, Routledge, London.

3 Attitudes to the Countryside

SALLY COOK, ADRIAN MOORE AND CLAIRE GUYER

Introduction

The late 1980s witnessed a massive and unprecedented rise in environmental awareness and public concern for environmental issues among the populations of the more affluent developed nations. Popular support for environmental and ecological conservation has been manifested by, among other things, a steep rise in the membership of environmental organisations, although it is clear that levels of public interest in these issues in Northern Ireland did not keep pace with those in the rest of the UK. There are particular features of life in Northern Ireland which may explain the lower level of environmental concern. Yearley (1995a) suggests that the decades of conflict have been detrimental to the fortunes of environmental issues in Northern Ireland, as immediate social and political concerns have taken precedence, while the sectarian nature of politics means that the larger political parties have seen no real gains to be made in campaigning on green issues.

In addition to these political factors, there are distinctive aspects of the rural environment which have probably helped to keep levels of public environmental interest below those in GB. In comparison with GB, Northern Ireland has a relatively high rural population, with a greater proportion in employment related to agriculture. On the whole farming activity is fairly small-scale and family based, and to a large extent the countryside of Northern Ireland has escaped the impacts of intensive farming commonly seen in much of lowland Britain. The relatively low level of industrialisation and the small scale of natural resource exploitation (for instance, Northern Ireland has little in the way of fossil fuel extraction apart from that of peat) means that environmental pollution has not been a widespread problem. Such factors have, according to Stringer (1992), led the Northern Irish population to adopt a far more sanguine approach to environmental problems in comparison with their British counterparts, and to maintain a lower level of concern for countryside issues.

The significance of agriculture and rural issues to the economy and the culture of Northern Ireland make it appropriate to undertake separate analysis of attitudes to the countryside, as distinct from those towards general environmental issues. Whereas the term 'environment' may conjure up images of a wide variety of issues, ranging from those with a purely local impact to those which operate on a global scale, the idea of 'countryside' is far more specific and more tangible, relating on the whole to a very restricted range of area types, settlement patterns and land uses. Unlike the concept of the environment, which may, to most people, be a rather abstract one (perhaps because it encompasses such variety), the countryside can be visualised and identified with more readily. Different people will, of course, maintain different conceptualisations of countryside, dependent perhaps on past experiences, on whether they work in the countryside, on whether they use the country as a place of recreation and relaxation - and particularly, on whether they actually live in a rural environment.

This chapter will begin with a general review of attitudes to the countryside in Northern Ireland, utilising comparative survey data to examine the current differences between Northern Ireland and GB, and to assess the extent of change in attitudes over recent years. Analysis will subsequently focus on variations in attitude according to the type of settlement in which the respondent lives. This is followed by analysis of associations between attitudes and social factors, aspects of economic activity and the frequency of recreational activity in the countryside. The final section attempts to determine whether apparent variations by residential area are explained simply by socio-economic factors, or whether area itself has an independent effect.

Attitudes to Countryside Protection in Northern Ireland

The persistence of a lower level of concern for the countryside among the Northern Irish population, as demonstrated in previous years by both Stringer (1992) and Yearley (1995b), is again apparent in Table 3.1. A higher proportion of respondents in GB is very concerned about things that may happen to the countryside, and while there is only a marginal difference in the proportions of those in Northern Ireland and GB who believe the countryside has changed over the last 20 years, those in Northern Ireland are substantially less likely to believe the changes to have been detrimental.

Stringer (1992) also points out that there is a tendency for new ideas to be taken up more rapidly by some groups than others, and that it is generally observed that the more affluent, more educated, and often more metropolitan population groups are the first to adopt new attitudes and behaviour; subsequently there is a lag time while new ideas 'trickle down' to people in poorer socio-economic circumstances, to those with a lower level of educational attainment, or to those in more rural, less developed areas. He speculated that attitudes to environmental and countryside issues in Northern Ireland would, after some lag time, tend to follow and even converge on those held by the British population. Indeed, data from the 1993 survey did demonstrate that the gap between GB and Northern Ireland with respect to environmental activism was closing (Yearley, 1995b).

Table 3.1   Concern about the countryside

	NI	GB
	%	%
Very concerned about the countryside	32	43
Countryside is much the same as 20 years ago	18	22
Countryside has changed for the worse	33	47

Table 3.2 shows firstly that attitudes have on the whole remained a little less protectionist in Northern Ireland, and, with the exception of one potential threat, the percentage of respondents in Northern Ireland favouring development over protection is between 2 per cent and 10 per cent higher than in GB. The exception to this, namely the provision of more picnic and camping sites in the countryside, is evidently seen as a relatively minor threat in any case, with nearly half the respondents in both surveys expressing encouragement for increasing the number of sites, and only around 3 per cent believing that it should be stopped altogether.

The biggest differences in attitudes relate to the provision of new roads and the building of new housing in country areas, both of which may be seen as being of particular benefit to those who live in the countryside; the accessibility of various types of service, and the availability of affordable housing, are often found to be particular problems for large sections of the rural population in many parts of the UK (Champion and Watkins, 1991). The differences between GB and Northern Ireland in relation to these issues may be partly explained simply by the lower level of urbanisation in Northern Ireland, and the consequent higher representation of rural residents in the surveys - 20 per cent of the Northern Irish sample lives in the countryside, compared to just 4 per cent of the British sample.   The relaxation of planning restrictions which took place from 1979 appears to be in keeping with the mood of the Northern Ireland population, among which, according to Christie (1996), there is widespread resistance to the restriction of rights and freedoms. Despite the detrimental environmental impacts (such as landscape degradation and damage to wildlife habitats) which have been a consequence of Green Belt and rural housing development, only 10 per cent of Northern Irish respondents believe that new building should be stopped altogether, and nearly half either don't mind or believe such development should be encouraged, in comparison to just under 30 per cent of British respondents.

Table 3.2   Protection or development of the countryside

	1987	1990		1996	
	GB	NI	GB	NI	GB
	%	%	%	%	%
Industry should be prevented from causing damage to countryside (even if higher prices)	83	82	89	91	93
The countryside should be protected from development (even if fewer jobs)	60	67	73	72	78
Should encourage:					
Increasing the amount of countryside being farmed	10	24	22	22	19
Putting needs of farmers before wildlife	9	12	6	8	6
Building new housing in country areas	12	15	8	12	7
Provision of more roads in country areas	20	20	10	18	8
More picnic areas and camping sites	54	47	47	47	48

The degree of similarity in the percentages of respondents who would support agricultural development is, perhaps, a little surprising in view of Stringer's (1992) comments in relation to the significance of agriculture in employment, and the political influence which the farming lobby wields. The farming sector in Northern Ireland is around three times as important economically as it is for the UK as a whole (DANI, 1995), and the percentage of the workforce employed in agriculture is, at 8 per cent, four times higher. There is, however, a potential ambiguity in the way in which the environmental impact of farming is perceived. While agricultural expansion may generally be seen as potentially harmful through the habitat damage or environmental pollution which can accompany agricultural practices, there may be circumstances in which farming can be seen as a more benign option to alternative land uses, particularly when these may include disruptive and highly visible industries such as mineral extraction or open-cast mining. It may be that British respondents from more industrial areas would favour agricultural development more than those living in areas dominated by intensive farming. Certainly, the position of farming seems to have been redeemed to some extent in the eyes of the British survey participants in the period between 1987 and 1990. While farming practices in Northern Ireland differ in nature and extent to those of Britain, being generally smaller-scale and less intensive, the region has by no means escaped environmental damage, especially in terms of habitat and species loss (Christie, 1996). It is possible that an awareness of such issues has had an impact on opinions in Northern Ireland, keeping them more in line with those of GB, where opposition to farming may be mitigated by different perceptions of the consequent environmental impact.

In relation to the 'trickle down' of environmental protectionism (Stringer, 1992), there are indeed signs of convergence in attitudes between Northern Ireland and GB on some countryside issues, although this may be due more to a slowing down in the rate of change in opinions which was observed during the late 1980s in Britain, rather than a more rapid progression to protectionist attitudes in Northern Ireland. For example, between 1987 and 1990, the percentage of respondents in GB believing that industry should be prevented from damaging the countryside increased by 6 per cent (from 83 per cent to 89 per cent), while in the following six years the additional increase was just 4 per cent. Changes in the proportion of those opposed to countryside development show a more extreme slow-down, from an increase of 13 per cent during 1987-1990 to 5 per cent in the subsequent six years. However, these observations are hardly surprising considering that by 1989 (a peak period for public support of environmental issues) such an overwhelming majority of people already favoured these forms of protection, and it is only to be expected that some limit, falling short of unanimous support, will exist in the weight of opinion favouring protection. Likewise, the percentage of British respondents who wished to encourage potentially harmful measures such as increasing housing, roads and picnic or camping sites, and putting the needs of farmers before those of wildlife, declined considerably more in the three years from 1987 than in the following six years. The last six years has seen a greater decrease in the percentage of those who could encourage rural development of new housing and putting the needs of farmers before those of wildlife in Northern Ireland than in GB, and the gap between beliefs on prevention of industrial damage has almost been closed in this period.

Table 3.3   Greatest threat to the countryside

	NI %	GB %
Land and air pollution, or discharges into rivers and lakes	34	28
Litter and fly-tipping of rubbish	26	14
New housing and urban sprawl	18	15
Industrial development (eg factories, quarries, power stations)	8	10
Building new roads and motorways	4	17
Superstores and out-of-town shopping centres	3	5
Changes to traditional ways of farming and of using farmland	3	3
Changes to the ordinary natural appearance of the countryside	2	2
The number of tourists and visitors	1	2
Other answer	1	1
None of these	1	1

Perceptions regarding the types of development or hazard which pose the most serious threat to the countryside differ a little between Northern Ireland and GB, although the rank-ordering of these is generally similar (Table 3.3). Notable differences are apparent, however, in relation to two potential threats, both of which also emerged as the exceptions to the general consensus in 1990 (Stringer, 1992). The first of these is the greater importance given by Northern Ireland respondents to the problem of litter and fly-tipping (which could be considered as one of the most easily remedied problems), deemed as the greatest threat by just over a quarter of the sample, compared to 14 per cent in GB.

By contrast, British respondents are four times more likely than their Northern Irish counterparts to consider that the construction of roads and motorways is the most important threat to the countryside - in fact this is the second most common response in the GB sample, but only ranks fifth in Northern Ireland. Although this question cannot be compared directly to Stringer's (1992) results as the categorisation of possible responses has changed since the 1990 survey, it is interesting to note that the percentage believing road-building to be the greatest threat has increased from 13 per cent to 17 per cent in Britain, but has actually declined from 8 per cent to 4 per cent in Northern Ireland. A low level of concern on this issue is, according to Stringer, explicable simply by the relative lack of road-building in Northern Ireland, and is consistent with the greater encouragement accorded to road construction, seen in Table 3.2. By contrast, there seems to have been a general increase in opposition to road-building in Britain (Yearley, 1995b), and recent road-building schemes which have achieved a high profile through well publicised environmental protests (such as the Newbury by-pass) may well have hardened British attitudes on this subject.

Variation in Attitudes by Area in Residence

Of particular interest in a review of attitudes towards the protection and development of the countryside is the variation in opinion between groups according to their place of residence. The question is posed as to whether views on these issues differ between those people who live in the countryside and who are therefore most directly affected by development or by protectionist measures, and those people who live in urban areas and who, while perhaps having purely altruistic reasons for environmental protection, may actually experience the countryside primarily as a source of recreation.

A second point of interest is the direction in which area of residence may bias attitudes. On the one hand, it could be suggested that those who live in an urban environment may be less concerned about protection because countryside issues have less immediate and less frequent impacts on them, while countryside residents may have greater empathy with, and concern for, the conditions of the environment in which they live. On the other hand, those living in urban areas do not have to suffer the potential inconvenience and restrictions of countryside protection. It could therefore be speculated that rural residents might adopt a more pragmatic approach to the countryside, and perhaps be willing to favour particular forms of development for the potential benefits that may accrue economically or in terms of housing and transport, even if these have an adverse environmental impact.

Survey participants classify themselves as resident in one of five categories, namely a big city, the suburbs of a big city, a small city or town, a country village, and open countryside. In fact, many of the attitudes recorded in the survey do show considerable variation by place of residence, although the breakdown of attitudes is not as straightforward as the foregoing hypotheses may suggest. Looking first at the attitudes expressed with regard to changes to the countryside (Table 3.4), it is evident that a far higher proportion of urban respondents claim not to be particularly concerned about such changes, and that this proportion decreases steadily with settlement size. The pattern is less consistent with respect to the degree of change over the last 20 years, although a smaller proportion of those in large settlements believe that there has been a lot of change.

Table 3.4   Countryside change by area of residence

	city %	suburbs %	town %	village %	country %
Not very concerned about the countryside	39	39	31	26	21
Countryside has changed a lot in last 20 years	47	47	52	64	55
Generally changes have been for the worse	22	36	34	34	31

In relation to the impact of these changes, however, there is no clear pattern in responses according to area of residence, although an increasingly pessimistic attitude appears to prevail in larger settlements, with the notable exception of those living in big cities who are far less likely to believe that changes have been detrimental. Therefore while respondents in country areas are more likely than those in towns or cities to be concerned about theoretical or potential changes, they are less likely than most other groups to believe that recent changes have in fact been harmful.

The relative lack of concern of city dwellers is reinforced in Table 3.5; 78 per cent believed that industry should be prevented from damaging the countryside even if this results in higher prices, compared to 91 per cent for the sample as a whole, and only 61 per cent would prefer countryside protection over employment. Interestingly, the proportions believing that the countryside should be protected even if this leads to lower employment are the same in villages as in the big cities, and in fact 37 per cent of villagers stated a preference for job creation even at the expense of damage to the countryside. Conservationist sentiments are expressed most frequently by those in small cities or towns, and while country residents also appear to favour protection in some circumstances, the response is more ambivalent when employment potential is at stake.

Table 3.5 Countryside protection versus development by area of residence

	city	suburbs	town	village	country
	%	%	%	%	%
Industry should be prevented from causing damage	78	91	94	86	93
The countryside should be protected from development	61	73	78	61	71

Area of residence appears generally to have little effect on beliefs regarding the greatest threat to the countryside, and there is no consistent pattern to responses by settlement size (Table 3.6).

Table 3.6 Greatest threat to the countryside by area of residence

	city	suburbs	town	village	country
	%	%	%	%	%
Litter and fly-tipping of rubbish	26	22	25	27	30
New housing and urban sprawl	16	20	18	22	12
Industrial development	10	10	8	2	11
Changes to appearance of countryside	6	2	2	2	2

It is worth noting, however, that respondents in the country are the most concerned about litter and fly-tipping, almost as many selecting this response as did land and air pollution, while residents of big cities are the group which is most concerned about changes to the natural appearance of the countryside. It should be added, however, that those in villages or open countryside were most likely to believe this to be the second most significant threat (this data is not tabulated).

Particularly interesting is the contrast in attitudes between those living in country villages, and those in the countryside itself, for two of the categories. Construction of new housing is most likely to be perceived as a significant problem by villagers, and least likely by countryside residents. By contrast, the latter group are far more likely to be concerned about industrial development than villagers.

While there is a strong, and statistically highly significant, division by area in views on whether farming activity should be increased[1] ($p<0.001$), there is no simple trend apparent (Table 3.7). City residents have the least interest in this issue, with equal proportions of people believing it should be encouraged as discouraged, but overall more than 70 per cent either state that they don't mind, or do not answer the question.

Table 3.7   Increasing the amount of countryside being farmed by area of residence

	city %	suburbs %	town %	village %	country %
Should be stopped / discouraged	14	33	26	18	30
Don't mind	54	48	49	63	34
Should be encouraged	14	17	22	14	35
Not answered	17	2	2	5	1

Villagers follow a similar pattern of response, while those living in big city suburbs are most opposed to this form of development - even so, half the respondents in this group do not hold any particular view. Those who actually live in the countryside have the strongest views on this issue, with only just over a third of respondents not expressing an opinion either way; and while they are comprised of a group which is by far the most encouraging towards agricultural expansion, at the same time a high proportion of country respondents is also opposed to it. However, the support for agricultural growth which is indicated does not extend to sacrificing wildlife protection - around two-thirds of respondents in all area types (except for big cities) believe that the needs of farmers should not be put before wildlife, and while those in the country are the most likely to place highest priority on farmers' needs, the proportion doing so is relatively low, at 14 per cent. Again, city residents have the lowest level of interest in this issue, with over half not expressing a particular viewpoint.

Although villagers were marginally more likely than other groups to perceive new housing as the greatest threat to the countryside, the proportion which actually opposes new housing development is lower than that of people living in the city suburbs, or in small cities and towns (Table 3.8). Country residents would give the most support to building new housing, with 29 per cent encouraging this form of development, 20 per cent higher than any other group. There is something of a tradition amongst rural families, perhaps even regarded as a right, of providing land to build houses for children of the family, and certainly it would be expected that people actually living in the countryside would have the most to gain from the availability of more housing. The viewpoint of urban residents on the other hand is probably influenced more by the potential impact on landscape and diminishing of recreational value.

Table 3.8 Building new housing in country areas by area of residence

	city %	suburbs %	town %	village %	country %
Should be stopped / discouraged	46	52	58	49	43
Don't mind	37	38	32	41	27
Should be encouraged	9	7	9	7	29
Not answered	9	4	2	2	1

Attitudes to the provision of more roads in the countryside vary relatively little by area of residence. Again, however, city residents are the most likely to have no strong view one way or the other, with 44 per cent of those in big cities or the city suburbs responding in this way. The strongest encouragement for road-building comes from those living in the countryside itself, at 26 per cent (compared to 14-17 per cent of the other groups), again suggesting a slightly greater degree of pragmatism in response to the problems experienced by those actually living in the country.

As previously pointed out, the issue of increasing the number of picnic areas and camping sites seems to be perceived as relatively trivial in terms of environmental impact. There is, however, a significant degree of variation in attitudes to this issue by area of residence ($p<0.05$). Unlike some of the other forms of development those living in the countryside or in small settlements are more likely to oppose it - between 10 per cent and 16 per cent of those in towns or cities believe that increasing the number of such sites should be discouraged, rising to 23-24 per cent of rural residents. It is unlikely that the latter respondents perceive these developments to be a serious environmental threat, although they may be more likely to see them as irrelevant to their own interests, and the responses may indicate a degree of antipathy towards visitors. In contrast, it is self-evident that urban residents have the most to gain from an increased number of such amenities because of the greater potential choice of recreation.

The data presented in this section has shown that attitudes towards countryside issues vary significantly according to settlement size and the type of environment in which the respondent lives, with a distinct rural/urban split in responses evident for some of the variables. Three features of the data should be reiterated: firstly, the level of interest of those in big cities is generally relatively low, with far higher proportions consistently choosing responses indicating that no strong opinions are held in this context. Secondly, in some cases the responses of those in villages are more like those of cities, and at other times accord more with those of the countryside. Thirdly, attitudes of those in the country appeared to be quite polarised on a number of issues, which may be partly an outcome of there being a lower proportion of respondents with no strong opinion, but may also indicate a more factionalised community, comprised of significant proportions of people with opposing viewpoints.

There are indications that stated beliefs are underlain to some degree by self-interest - for example, countryside residents are far more likely to be in favour of measures which would be perceived as being of direct benefit to local communities, despite potentially adverse environmental impacts, while in general urban respondents tend to prefer the promotion of measures which would favour conservation of the existing environment. This may result from a lack of awareness of the problems of rural living, or may occur simply because they would like to preserve the amenity value of the countryside and do not have to suffer any negative impacts. An attempt will be made in the following section to test this speculation by examining the variation in attitudes according to the frequency of recreational activity in the countryside, although the data available is very limited. However, other explanations for the observed patterns must also be examined, since the area-based bias does not explain why, for instance, city residents have less concern for the condition of the countryside than those living in the suburbs, or why the views of villagers do not consistently concur with those of the countryside or of the larger settlements.

Impact of Socio-Economic Variables and Rural Recreation

Although the primary focus of this analysis is the effect of area-type on attitudes to the countryside, it is necessary at this point to examine whether the observed variation in attitudes can be explained by other features of the socio-economic, political or demographic profile of the survey sample in each area-type, rather than by area-based factors *per se*. The attitudes data were screened for associations with a number of variables, including the age-group, sex, social class and educational attainment of the respondent, the presence in the household of children, household income and political orientation. While strong associations were found between attitudes and socio-economic factors, in general very few significant differences were found between attitudes when compared by the sex of the respondent, the presence of children in the household, or political inclination. In a small number of cases attitudes were related to the age-group of the respondent, although this was not manifested in a consistent way. Subsequent analysis therefore concentrated on the impact of social factors, excluding those variables which showed weak or inconsistent associations for the sake of clarity. Social class and education level were both retained for this aspect of the analysis - while household income was often associated with environmental beliefs, the results were generally repetitive of the other two social variables used, and are not presented.

The variation in environmental beliefs according to education levels and socio-economic circumstances is well-documented, and has been investigated in previous analyses of attitudes in Northern Ireland (Stringer, 1992; Yearley, 1995b). The variation in social class (Table 3.9) is highly significant (p<0.001), with a steady increase in the proportion of professional and managerial classes from large settlements through to country areas. The proportion of semi-skilled and unskilled manual workers shows a comparable decrease in general, although country areas are intermediate between villages and towns with respect to the proportion of social classes IV and V, indicating that socially the countryside is marginally more polarised than the other areas. The breakdown of qualifications by area shows that cities have the lowest proportion of respondents with third level education and the highest proportion with no educational qualifications, while country areas have the greatest proportion of well-qualified people and the second lowest proportion of unqualified respondents; however, the division of qualifications by area is less marked than that of class.

Table 3.9   Social class and education by area of residence

	city %	suburbs %	town %	village %	country %
Social class I and II	17	24	28	39	42
Social class IV and V	38	34	27	12	20
Third level education	14	21	20	20	24
No educational qualifications	58	43	35	40	37

The influence of class and education on environmental attitudes, together with the area-based division of social class, could therefore tend to produce a higher proportion of protectionist attitudes with decreasing settlement size. This has been partially evident in the data, particularly with reference to the responses of city residents, but has by no means been an overwhelming feature of the analysis. In order to examine more closely whether area itself has an independent effect on attitudes to the countryside, or whether observed variations are explained simply by the class divisions, it is first necessary to determine the magnitude and pattern of difference in attitude by socio-economic characteristics.

For the purpose of the following analysis, social class is divided into four categories (social classes I and II combined; III non-manual and III manual as separate categories; and class IV and V combined), and educational attainment into five categories (namely degree or higher; third-level sub-degree qualification; 'A' levels; 'O' levels and GCE; and no qualifications).

Table 3.10   Concern about the countryside by highest educational qualification

	degree %	sub-degree %	'A' level %	'O' level %	none %
Not concerned about the countryside	6	22	23	39	37
Much the same as 20 years ago	17	13	21	18	17
Countryside has changed for the worse	49	34	34	35	29

There is a strong degree of variation in the frequency of concern for the countryside according to both social and education level; 42 per cent of those in social classes IV and V state that they are not particularly concerned, compared to 15 per cent of those in social classes I and II, and only 6 per cent of those with degree level education give this response (Table 3.10).

Beliefs regarding the amount of change in the last 20 years differ little by these social variables, but of those who do think the countryside has changed, social class I and II and people educated to degree level are more likely to believe that these changes have been detrimental, with views at the extreme ends of each categorisation being substantially different from the intermediate groups.

Social class and education level also appear to have an impact on attitudes towards preventing industry from causing damage to the countryside, although the division in responses is perhaps not as extreme as would be expected considering that the stated corollary of this measure is a possible increase in prices - 97 per cent of social classes I and II opted for protection of the countryside, compared to 86 per cent of those in the manual classes, while between 96 per cent and 100 per cent of those with post-'O' level qualifications maintained this opinion, compared to 86-88 per cent of those with lower (or no) qualifications.   Similar social class differentials were evident in response to a choice between countryside protection or job creation; 75-78 per cent of non-manual workers elected for countryside protection, compared to 64 per cent of those in social classes IV and V.   However, the response by education levels is far less straight-forward, with those least and most qualified showing similar percentages in support of job creation, at 28-29 per cent, compared to 18-19 per cent among the other groups.

It was noted (Table 3.5) that a particularly low proportion of villagers support countryside protection at the expense of job creation, and it could be speculated that this may be related to unemployment levels in villages - 8 per cent of respondents in this group are registered unemployed, compared to 7 per cent in towns, and 4 per cent in each of the other area types. Interestingly, however, the response to this question by economic activity was in part counter-intuitive; while it could be expected that people would relate these questions to their own personal circumstances, and therefore have a different perception of the potential economic drawbacks of environmental protection, in fact the data shows that a greater proportion of those in employment favour job creation (at 26 per cent) than do those currently unemployed (18 per cent).   Respondents in full-time education were the group which most commonly preferred the option of job creation even at the expense of possible damage to the countryside, with 39 per cent choosing this option. Thus while the unemployed do not display a strong tendency towards self-interest in choosing greater employment over protection, those who presumably intend to enter the job market on completion of their education do to a much greater extent, and the attitudes of this group (61 per cent of whom already qualified to at least 'A' level standard) are more heavily weighed in favour of job creation than any other single group examined (whether split by area, social class or education). It must, however, be pointed out that these observations regarding the unemployed and those currently in education are based on small sub-samples and may therefore be unreliable.

Table 3.11 shows no clear gradient in the choice of greatest threat to the countryside when divided by social class, although manual groups are more likely to select litter and fly-tipping, and less likely to choose housing or industrial development than non-manual workers.

Table 3.11   Greatest threat to the countryside by social class

| | I + II | III nm | III m | IV + V |
	%	%	%	%
Litter and fly-tipping of rubbish	22	21	30	28
New housing and urban sprawl	21	20	13	17
Industrial development	12	10	5	6
Changes to appearance of countryside	2	2	3	3

The breakdown of opinions by education level (Table 3.12) is far less straightforward, although again both litter and fly-tipping and industrial development may be picked out as categories displaying a wide range of responses at the extremes of qualification.

Most of the specific issues of countryside development show some differences according to the social class or education level of the respondent. Variation in attitudes to the increase of farming, however, is not statistically significant by social class and shows no particular trend, and while the most highly educated category is the most protectionist in this respect, there is no consistent pattern among the other groups. However, clearer conservationist trends are apparent by both class and education level in choosing between the needs of the farmer or those of wildlife; 71 per cent of social classes I and II oppose putting priority on farmers' needs, compared to 60 per cent, 66 per cent and 57 per cent of classes III non-manual, III manual and classes IV and V respectively.

Table 3.12   Greatest threat to the countryside by highest educational qualification

	degree %	sub-degree %	'A' level %	'O' level %	none %
Litter and fly-tipping of rubbish	14	25	14	23	34
New housing and urban sprawl	20	23	23	18	14
Industrial development	16	9	12	9	5
Changes to appearance of countryside	3	2	4	2	2

Professional and managerial workers are most likely to believe that construction of new housing in the countryside should be stopped or discouraged (at 65 per cent compared to 43-50 per cent among the other social classes), and social classes IV and V are marginally more likely to encourage it (15 per cent compared to 11 per cent of other classes). The division of answers is more extreme by education levels (Table 3.13). Considering the paucity of low-priced new housing, and the tendency towards the gentrification of the countryside, it is a little ironic that the people most likely to be able to afford such new housing are apparently most likely to oppose its construction.

Table 3.13   Building new houses in country areas by highest educational qualification

	degree %	sub-degree %	'A' level %	'O' level %	none %
Should be stopped / discouraged	75	59	58	51	42
Should be encouraged	2	11	4	14	17

Similarly, there is a clear trend by social class in the proportion of respondents who believe that the provision of new roads in country areas should be stopped or discouraged, with percentages varying between 57 per cent and 37 per cent for the professional/managerial and the semi-skilled/unskilled categories respectively. Responses by education level are more mixed, although again there is a tendency for those at the extreme ends of the spectrum to have the greatest differences of opinion. However, views on increasing the number of picnic areas and camping sites do not differ significantly by social class, but the division of response by education tends towards a bimodal split, with those most and least qualified offering similar attitudes.

Finally, a brief look can be taken at the crosstabulation of attitudes against the frequency of recreational use of the countryside. Information on this is limited to a question which asks whether the respondent has, in the past year, visited the countryside or the coast for an outing of some kind; responses are coded as none, one or at least two outings. Attitudes to most variables follow a pattern whereby those who use the countryside for recreational purposes are more likely to have concern for rural issues - for example, nearly three-quarters of those who have been on at least two outings expressed some concern for things that might happen to the countryside, compared to just over half of those who had used the countryside for one excursion, and just over 40 per cent of those who had not been on any outings; the proportions believing that industrial damage should be prevented were 93 per cent, 87 per cent and 74 per cent for the above categories. However, it is difficult to speculate from this whether the degree to which the countryside is used for recreation directly influences respondents' views, since the frequency of recreational use also varies significantly by social factors - only 7 per cent of professional/managerial workers had not been on a country outing in the previous year, compared to 15 per cent of semi-skilled and unskilled groups.

Relative Effects of Area and Social Factors

The relative importance of area and social factors to attitudes can be explored a little further by generating three-way crosstabulations involving these variables. The data can then be compared by class for each area type (to determine whether the effect of class is consistent in different environments) and for each class between different area types (to discover whether area of residence has a differential effect on those in the same social grouping).

Table 3.14   Not very concerned about the countryside by area of residence and social class

	non-manual	manual
	%	%
City	28	45
Suburbs	31	44
Town	26	38
Village	18	38
Country	19	28

An examination of those expressing a lack of concern for the countryside divided by social class, calculated separately for each area type (Table 3.14) shows that non-manual respondents in all settlement types are consistently less likely to say they are not concerned, and furthermore that the tendency for fewer unconcerned responses with decreasing settlement size applies to both manual and non-manual social groupings. It should be particularly noted that the smallest difference between the social groups is found in the country areas.

Attitudes to the construction of new housing is similarly crosstabulated (Table 3.15). The percentage of non-manual respondents opposed to new construction is surprisingly similar across the range of area types, varying by a maximum of 11 per cent, although it is still the case that members of this social grouping that live in the country are less likely to discourage house-building than their social counterparts in other areas. Far bigger differences are observed for responses among the manual social group, the largest difference being that between the suburbs and the countryside, at 37 per cent. In fact 47 per cent of manual respondents in the countryside believe that new building should be encouraged, compared to between 5 per cent and 11 per cent of respondents in the same social group living in other areas. While it is clear that both social class and area of residence influence attitudes on this issue, the two factors do not interact in a straightforward way, but rather seem to have a disproportionate impact on manual respondents living in the countryside.

Table 3.15   House-building should be stopped or discouraged by area
of residence and social class

	non-manual	manual
	%	%
City	60	36
Suburbs	56	55
Town	64	49
Village	56	41
Country	53	18

The variable recording the frequency of recreational activity in the countryside can also be examined through three-way crosstabulation of the data, to allow a more detailed exploration of the separate effects of recreation, area and social factors.   Firstly, while recreational use is consistently associated with greater concern for the countryside, the amount of difference this makes to attitudes is far less outside the larger settlements, especially in the countryside itself, than in the towns and cities (Table 3.16). While countryside respondents are less likely to express concern if they have been on no outings in the previous year, the percentage doing to is, at only 29 per cent, similar to that of urban respondents who have been on two or more excursions.

Table 3.16   Not very concerned about the countryside by area of
residence and number of outings in previous year

	none	one	two or more
	%	%	%
City	70	44	26
Suburbs	72	40	35
Town	63	65	27
Village	58	25	23
Country	29	26	19

A similar pattern is found with respect to the prevention of industrial damage: while only 44 per cent of those in cities who had been on no outings in the previous year believe that industry should be prevented from damaging the countryside (compared to 88 per cent of city residents who had been on at least two excursions), the number of outings makes no difference to the response of country residents, 92-93 per cent of whom believe that industrial damage should be prevented.

Even within each social grouping, attitudes are clearly graded according to the frequency of recreational use. Table 3.17 looks at the lack of concern for the countryside by the number of outings among manual and non-manual social groups. While a lower level of concern among manual respondents is maintained even among those who do use the countryside for recreation, nevertheless this group is only half as likely to say that they are not particularly concerned as are those in the same social group who have had no excursions in the previous year.

These trends, which emerge even when the sample is split into four social groups, imply that the apparent influence of recreational use of the countryside on attitudes is not a function of social class alone.

Table 3.17 **Not very concerned about the countryside by social class and number of outings in previous year**

	none %	one %	two or more %
Non-manual	52	30	22
Manual	68	54	32

It would be interesting to examine the impact of agricultural employment on attitudes to countryside and farming issues, but unfortunately the sample size of farm-related workers is too small for any type of statistical viability; however, tentative observations may be made. Selecting only respondents living in the countryside, it is found that a far higher proportion of agricultural workers favour job creation over countryside protection, that they are more likely to believe that expansion of farming should be encouraged (50 per cent compared to 31 per cent of non-farming country residents) and that the needs of the farmer should come before those of wildlife. They are also more likely to support the construction of roads (47 per cent believing that provision of new roads should be encouraged, compared to 23 per cent). A far higher proportion perceives litter and fly-tipping as the most serious threat to the countryside, and fewer see urban sprawl and environmental pollution as significant problems. The difference in attitudes between farmers and non-farmers could in part explain the polarisation of views among country residents observed for several variables.

Conclusion

This chapter has focused on a small selection of specific issues related to countryside protection and development. Although there is as yet limited potential for tracing the progress over time of attitudes towards the countryside in Northern Ireland, the data appears to confirm that the frequency of protectionist attitudes is increasing, and that in most respects the gap between Northern Ireland and GB is slowly but surely being closed. However, respondents in Northern Ireland remain considerably more likely than their British counterparts to favour certain forms of countryside development, especially those which would be expected to produce tangible benefits for rural communities, namely the construction of new housing and new roads in country areas.

Examination of Northern Irish attitudes according to the type of area in which the respondent lives, and the social class to which he or she belongs, highlights the fact that strong differences of opinion exist between population sub-groups with respect to a number of issues. While people are more likely to claim to be concerned about the countryside if they actually live in a rural environment, at the same time urban respondents in most areas (towns, small cities and big city suburbs) are somewhat more inclined to express protectionist attitudes, particularly in relation to house- and road-building. A relatively high proportion of those living in big cities do not maintain any particular opinion on a number of countryside issues, whereas those resident in the country are more likely than other groups to express a strong opinion; however, the split in responses of this latter group indicates a greater lack of consensus on certain issues within the rural community. The apparent polarisation of views may be partially explained by the fact that those people in agricultural occupations are far more likely than other countryside respondents to be in favour of developmental measures. Indeed, it is likely that a significant proportion of rural residents who are not currently engaged in agriculture, but who come from a farming background, still hold attitudes more supportive of farmers' interests than do rural immigrants, who may have had little family connection with farming or even with the countryside.

Social class and education level are strongly associated with support for many aspects of countryside protection. As one of the more extreme examples, people with no educational qualifications are six times more likely to state that they are not particularly concerned about the countryside than those educated to degree level, although in general there is at most a two-fold difference in protectionist attitudes at the extremes of education or social class.

However, while the proportions in each social class vary widely between area types, with the highest proportion of semi-skilled and unskilled respondents living in the cities, and professional and managerial respondents most commonly found in country areas, this does not explain the observed area-based variation in attitudes. Exploration of attitudes within each social grouping suggests that differences do exist between those living in large settlements and those living in country areas, which are not underlain by the different socio-economic profiles of each environment. The case of house-building illustrates well the gap in attitudes between social classes, which is maintained within most area types, but more significantly in this context it also implies that area of residence may have a direct impact on attitudes. While the influence of area on the non-manual group is relatively minor, it appears to have a disproportionate impact on the views of the manual group.

It is therefore concluded that the settlement type in which a person is resident does have an independent effect on attitudes to many countryside issues, although it is not possible to characterise either urban or rural populations as having more interest in the countryside and its conservation. While countryside respondents are far more likely to claim to be concerned about the countryside, they are also, on the whole, more likely to favour some forms of countryside development such as the construction of new houses. Interestingly, the preference for development doesn't apply to the provision of more recreational areas (camping and picnic sites), which are opposed by considerably more of those living in the country than by those in urban areas. While it is not likely that such sites are regarded as an environmental threat, they may be seen as entirely irrelevant to the needs of the rural population.

Although urban residents are more likely to resist countryside development (aside from the provision of picnic and camping sites), it cannot be said with certainty that this is entirely motivated by a genuinely altruistic commitment to conservation and environmental protection. After all, it could be said that a significant proportion of countryside respondents display a pragmatic self-interest with regard to some of the development/protection issues, and it could be speculated that urban respondents have revealed a similar level of self-interest - perhaps they are more in favour of conservationist measures since they are in a position to exploit the recreational and amenity facilities of the countryside without having to suffer the potential inconveniences of rural life. The fact that the proportion of respondents expressing conservationist sentiments was strongly associated with the frequency of rural excursions, even when social class was held constant, certainly does not contradict this conclusion.

Notes

1. All statistical tests referred to in this chapter were chi-square tests.

References

Champion, T. and Watkins, C. (1991), 'Introduction: Recent Developments in the Social Geography of Rural Britain', in T. Champion, and C. Watkins (eds), *People in the Countryside: Studies of Social Change in Rural Britain*, Paul Chapman, London.

Christie, S. (1996), *Environmental Strategy for Northern Ireland*, Northern Ireland Environment Link, Belfast.

DANI. (1995), *An Overview of the Northern Ireland Agri-Food Industry*, DANI, Belfast.

Stringer, P. (1992), 'Environmental Concern', in P. Stringer, and G. Robinson, (eds), *Social Attitudes in Northern Ireland: The Second Report*, Blackstaff, Belfast.

Yearley, S. (1995a), 'The Social Shaping of the Environment Movement in Ireland', in P. Clancy, *et al* (eds), *Irish Society: Sociological Perspectives*, Institute of Public Administration, Dublin.

Yearley, S. (1995b), 'Environmental Attitudes in Northern Ireland', in R. Breen, P. Devine, and G. Robinson, (eds), *Social Attitudes in Northern Ireland: The Fourth Report*, Appletree Press, Belfast.

4 Role of Government

NIALL Ó DOCHARTAIGH

Northern Ireland: A Form of Government Rejected by Government

Any discussion of the role of government in Northern Ireland must address the fact that the very form of government in Northern Ireland is a hotly contested political issue. Virtually all political parties in Northern Ireland have regarded the system of government by direct rule from London since 1972, (with the addition of some involvement by the Irish Government since 1985) as unsatisfactory. This negative political consensus has extended from the advocates of full integration into the UK, through the mainstream unionist advocates of a devolved administration of one kind or another, to nationalists and republicans who have questioned the basic legitimacy of government in Northern Ireland. Both unionists and nationalists have complained that Northern Ireland is governed as a colony; in the case of one DUP politician, as a 'shared colony' of the British and Irish Governments (cited in O'Leary and McGarry, 1993, p.220). As Connolly and Erridge (1990, p.33) have noted, 'The structure of government in Northern Ireland has greatly strengthened the role of the civil service in the policy process' at the expense of elected politicians. Unionists in particular argue that there is a 'democratic deficit' in Northern Ireland, implicitly suggesting that it can be resolved through the reform of structures of government within Northern Ireland. There has been deep disagreement on the appropriate form of government but there has nonetheless been a consensus among the political parties that the *status quo* was unacceptable. This consensus has also extended to those responsible for government in Northern Ireland over the past decades. While certain British administrations tacitly regarded direct rule as the best option for governing Northern Ireland because it at least provided stability, virtually all British administrations have made some attempt to change the form of government in Northern Ireland. The plethora of initiatives which have been launched since 1972, if they demonstrate nothing else, demonstrate that in Northern Ireland government itself sees the current form of government as fundamentally unsatisfactory.

Within this negative consensus there are two directly opposing viewpoints on the level at which change in government needs to take place. Unionists argue that the focus should be on changes to the form of government in Northern Ireland, changes at the 'institutional level'. Nationalists argue that the constitutional framework for government in Northern Ireland has to be addressed and changes made at the 'constitutional level' (Todd, 1990, pp.3-7). At the core of debate over government is a disagreement on the level at which this issue needs to be addressed. In the mid and late 1990s the British and Irish Governments embarked on a 'three-stranded' peace process which dealt with forms of government in Northern Ireland in tandem with relationships between the two parts of Ireland and between Ireland and Britain. In doing this they implicitly accepted that the issue of forms of government within Northern Ireland was not a solely internal matter to be resolved in isolation from relationships between the two parts of Ireland. Despite this, there is in the peace process a constant tension between the two views.

The business of governing Northern Ireland has been conducted by successive direct rule administrations for the past quarter of a century. This form of government has brought a degree of political stability and, unsatisfactory though it may be, has been the *status quo* for a quarter of a century and derives immense authority from that fact alone. The legitimacy of government is contested but a large proportion of the institutions of the state are accepted unproblematically by, and enjoy substantial loyalty from, even the most disaffected sections of the population, not least those institutions associated with health, education and social welfare. The workings and institutions of the modern state are so extensive and fundamental to human existence that it is unrealistic for a population of even the most disaffected to reject the state in all or even most of its forms.

Before looking directly at attitudes to government and the role of government in Northern Ireland it is useful to reflect on attitudes to politics in general in Northern Ireland and attitudes to Northern Ireland - the political framework within which government operates.

Attitudes to Politics: Lack of Interest and Bewilderment

Given Northern Ireland's reputation for political passion and intransigence it is more than a little ironic that a higher proportion of both men and women in Northern Ireland should declare themselves uninterested in politics than their counterparts in GB. In both GB and Northern Ireland women declare considerably less interest in politics than men.

Table 4.1   How interested would you say you personally are in politics? (by gender and religion)

	GB %	GB men %	GB women %	NI %	NI men %	NI women %	NI Prot. %	NI Cath. %	NI other %
Not very interested	24	20	27	28	25	31	27	30	28
Not at all interested	12	7	15	15	14	16	12	16	22
Total not interested	36	27	42	43	39	47	39	45	50

In GB the gap is striking, with 42 per cent of women declaring themselves not very or not at all interested in politics compared to 27 per cent of men. The gap between men and women is far narrower in Northern Ireland but this is largely because there are such low levels of interest among both sexes. The gap between men in GB and men in Northern Ireland is wide; 39 per cent of men in Northern Ireland say they are not very or not at all interested compared to 27 per cent of men in GB.

There is also a small but noticeable gap in attitudes between Catholics and Protestants in Northern Ireland with Catholics being more likely to be uninterested while the 'others' (those who have no religion, did not declare their religion or are members of small non-Christian minorities) are the least interested of all.

Thus, while women in GB are far less interested in politics than men in GB, revealing a worrying gender gap, they are still not as uninterested as Northern Irish Catholics are. These figures should at least unsettle the assumption that the population of Northern Ireland is obsessed with politics and that such obsession has been a major source of the conflict.

Part of the explanation for this lack of interest lies with the nature of politics in Northern Ireland in recent decades. Politics has been so mixed up with destruction and death that there has been a depoliticisation of large sections of the population - not least in areas with high levels of violence. For many people 'politics' has been a choice between the unacceptable and the unrealistic. Many people in Northern Ireland are also reluctant to express 'interest' in politics, as though it was a sport. In addition there has often been danger associated not only with active political participation, but even with the expression of political opinions. In certain situations, it has been physically dangerous to be 'interested' in politics.

The generally high levels of voting at elections in Northern Ireland suggests that, while people may not express interest in politics they participate in the political process nonetheless and at one of the highest rates in western Europe. Nonetheless the low levels of interest point to an extensive depoliticisation of the population (while of course certain sections have become very highly politicised). Contrary to some of the stereotypes, people in Northern Ireland tend to be that little bit less interested in politics than their counterparts in GB.

Table 4.2   Sometimes politics and government seem so complicated that a person like me cannot really understand what is going on

	GB %	GB men %	GB women %	NI %	NI men %	NI women %	NI Prot. %	NI Cath. %	NI other %
Agree strongly	22	16	27	27	23	30	26	27	27

It is worth noting at this stage that the relatively lower level of interest in politics among women is reflected in other areas of the survey where there is a marked tendency for women to cluster in the moderate middle.

The figures can be taken as a measure of general 'bewilderment' with politics and government. Table 4.2 shows only those who 'agree strongly' with the proposition and shows that if lack of interest is high (taking only those who claim to be not at all interested) bewilderment is far higher. Overall levels of 'bewilderment' using this measure appear higher in Northern Ireland (27 per cent) than in GB (22 per cent). The most striking finding is the different levels of bewilderment by gender. In both Northern Ireland and GB, but particularly in GB, women feel considerably more bewildered than men. Once again, the gender gap is narrower in Northern Ireland largely because of the generally high levels of confusion here. However, as Hayes and McAllister (1996, p.17) have pointed out, there are large variations in the levels of bewilderment between England, Scotland and Wales. While England and Wales had lower levels of bewilderment than Northern Ireland in 1994, Scotland had considerably higher levels.

The results of these two questions on attitudes to politics in general provide the context within which to assess the results outlined below. They suggest a lack of direction and a lack of interest among large sections of the population and a population more detached from and confused by politics than those in GB.

The Constitution of the State

The fact that there are surprisingly low levels of interest in and understanding of politics and government in Northern Ireland does not detract from the fact that there is fundamental disagreement on the existence of the state itself and substantial opposition to any government of Northern Ireland, no matter what form it takes.

Table 4.3 shows answers to a question which asked people how much they agreed or disagreed with the phrase 'I would like the future of Northern Ireland to be within a united Ireland'. It provides a very soft measure of support for a united Ireland measuring the outer limits of that support and corresponding to 'aspiration' rather than 'demand' or 'conviction'. It also provides a negative measure of support for the *status quo* by measuring opposition to a united Ireland. These figures can be seen as a crude and unsatisfactory but still useful measurement of fundamental attitudes to the existence of the state. It is not a very demanding question: people can support a united Ireland in theory without ruling out any other option and without considering a time frame apart from the dim and distant future. It shows that Catholic support for or 'aspiration' to a united Ireland is substantial and strong, but far from overwhelming even by this softest of measures. Only a slight majority of Catholics say they would like to see the future of Northern Ireland in a united Ireland. By contrast Protestants are far more in agreement on this issue with three quarters of Protestants opposing a united Ireland even by this soft measure.

The question posed in Table 4.3 provides a means of tentatively quantifying two much talked about groups - Catholic unionists (13 per cent of Catholics - those who disagree or strongly disagree with the statement) and Protestant nationalists (7 per cent of Protestants - those who agree or strongly agree with the statement). Even by this soft measure there are more of the former but the proportion of Protestants strongly favouring a united Ireland (4 per cent) is the same as the proportion of Catholics strongly opposed to it (4 per cent).

Table 4.3   **'I would like the future of Northern Ireland to be within a united Ireland'**

	NI %	NI Prot. %	NI Cath. %	NI other %
strongly agree	11	4	23	5
agree	14	3	30	13
neither	25	16	33	37
disagree	20	27	9	23
strongly disagree	30	51	4	23

In previous surveys from 1989 to 1994 respondents were offered a bald choice on the constitutional issue in a question which asked 'Do you think the long-term policy for Northern Ireland should be for it to remain part of the UK or to reunify with the rest of Ireland'. Generally about a quarter to a third of Catholics answered that Northern Ireland should remain in the UK and this question has been used previously as a tentative measure of the number of Catholic unionists (Breen, 1996, pp.35, 45-46). I would suggest that the figures above using the softer measure of support for a united Ireland give a better estimate of the number of 'Catholic unionists' in the sense of people who actually oppose a united Ireland to some degree. This is the group which might conceivably contribute to the maintenance of the Union at some stage by, perhaps, voting unionist. It is a much smaller group than previously suggested by the stark question in previous surveys offering the choice of united Ireland or UK.

One of the oddities of these figures is that, when cross-referenced with national identity, they show that those who strongly favour a united Ireland are more likely to identify themselves as Northern Irish than the population at large, and much more likely than those who oppose a united Ireland. This suggests that the development of a 'Northern Irish' identity, often seen as one building block of a settlement within Northern Ireland, does not mark a shift to acceptance or support for Northern Ireland as a state. It is entirely compatible with support for a united Ireland.

There is a further oddity about these figures and one that suggests a great deal of caution must be used when dealing with them. Of those who agreed that they would like the future of Northern Ireland to be within a united Ireland fully 4 per cent described themselves as unionists while of those who agreed strongly 12 per cent described themselves as unionists. This suggests that a small section of northern unionists, however vaguely, might still aspire to a reunification of Ireland within the UK, a return to the days before an independent Irish state, in a sense the ultimate unionist 'aspiration'. This is not an option which has been offered in surveys in recent years. However Whyte describes how in 1967, just before the outbreak of the Troubles, a Belfast Telegraph survey asked people which of the following arrangements they thought 'would be best for Ireland?': The 'situation as it exists today' (then a Stormont parliament); 'An independent united Ireland'; or a 'united Ireland linked to Britain'. In 1967 just over 40 per cent of Protestants chose a 'united Ireland linked to Britain' (Whyte 1990, p.77). This sense of the link to Britain rather than the partition of Ireland as the core of unionism has been greatly eroded over the course of the conflict but clearly there is still a small group of unionists who have some affection for the ultimate unionist ideal. One unionist politician could say in 1991, 'I would be quite prepared to take a united Ireland tomorrow, if somehow the whole of Ireland could have some form of union grafted [on]' (McGimpsey cited in Cochrane, 1997, p.37). It is tempting to speculate that this submerged aspect of unionism might come floating to the surface in a peace process which in theory will be dealing with links between the two parts of Ireland together with links between Ireland and Britain.

Perhaps the most important thing about these figures is that they show that Catholic opposition to a united Ireland, a key test of loyalty to the constitutional *status quo*, is extremely weak. Those Catholics not supportive of a united Ireland and by implication not opposed to Northern Ireland remaining in the UK, show minimal levels of loyalty to the constitutional *status quo*. The foundations of any form of Northern Ireland are shaky not merely because of the number of Catholics who support a united Ireland, but above all because the lack of support for a united Ireland by large numbers of Catholics shows little sign of converting to loyalty to or support for Northern Ireland staying within the UK.

Attitudes to Government: a Deep Distrust

Table 4.4 should be understood in the light of widespread bewilderment, lack of interest and a basic constitutional framework for government which has a weak support base and thus an uncertain future.

There are low levels of trust in Northern Ireland that any British Government will act in the best interests of Northern Ireland. In fact a greater proportion of the population of Northern Ireland distrust the British Government than distrust a putative united Ireland Government to act in the

best interests of Northern Ireland. Of course this question doesn't relate to
trust in the institutions of the state or trust in its ability to deal with day to
day matters. Undoubtedly it measures very different views of the best
interests of Northern Ireland dominant in the Protestant and Catholic
communities.

**Table 4.4   Under direct rule from Britain, as now, how much do you
generally trust British Governments of any party to act in
the best interests of Northern Ireland?**

	NI	NI Prot.	NI Cath.	NI other	GB
	%	%	%	%	%
Rarely	27	21	31	34	10
Never	14	11	18	15	3
Total not trusting	41	32	49	49	14

It has to be noted also that as the actual, rather than potential governing
agency in Northern Ireland the British Government is much more open to
distrust than putative united Ireland or Stormont Governments as it actually
makes the decisions in Northern Ireland (albeit since 1985 in some sort of
consultation with an Irish Government which might be seen to represent a
potential united Ireland Government). Having said all that, the levels of trust
for the direct rule government seem disastrously low and highlight the
problematic relationship between ruler and ruled under direct rule. It
suggests fairly low levels of identification with government by either
Catholics or Protestants. Catholics are far less distrustful of a potential
united Ireland Government than of the British Government while Protestants
are far less distrustful of a potential Stormont Government than of the
British Government. On the other hand Protestants are significantly more
distrusting of a united Ireland Government than they are of the British
Government. While a potential Stormont Government is usually seen as the
Catholic community's nightmare scenario, Catholics are almost exactly as
distrusting of the British Government as of a potential Stormont
Government. The British Government may be distrusted by both Catholics
and Protestants but it is by no means situated half way between the
affections (or disaffections) of the two communities. While there are high

levels of distrust by both communities, Catholics are notably more likely than Protestants to distrust the British Government.

Table 4.5   **If there was self-rule, how much do you think you would generally trust a Stormont Government to act in the best interests of Northern Ireland?**

	NI %	NI Prot. %	NI Cath. %	NI other %	GB %
Rarely	14	7	23	13	8
Never	15	7	27	15	3
Total not trusting	29	13	49	28	11

And if there was a united Ireland, how much do you think you would generally trust an Irish Government to act in the best interests of Northern Ireland?

	NI %	NI Prot. %	NI Cath. %	NI other %	GB %
Rarely	21	28	15	19	7
Never	17	27	6	16	2
Total not trusting	39	54	21	35	9

The GB sample showed extraordinarily low levels of distrust of either British or prospective Stormont or all-Ireland governments. It is probably fair to say that this is less illustrative of deep trust than it is of lack of interest.

The next question (Table 4.6) measures trust of government in a more general sense, not just as it relates to the uncertain future of Northern Ireland as a unit. It is taken here as a measure of trust of government and the state in general that people are quite content for government to hold large amounts of information about citizens. It measures the extent to which the institutions of the state and, implicitly, the security apparatus of the state are trusted to use their powers wisely and in the interests of the average citizen.

Table 4.6   The government has a lot of different pieces of information about people which computers can bring together very quickly. (How much of a threat to individual privacy is this...)

	NI %	NI Prot. %	NI Cath. %	NI other %	GB %
Very serious	30	21	42	30	28
Fairly serious	33	32	35	37	37
Not serious	31	39	19	33	30
Not a threat at all	6	9	5	*	5
Total saying it's not a serious threat or not at threat at all	37	48	24	34	36

By this measure Northern Irish Protestants are notably more trusting of government than the sample in GB while Catholics are notably less trusting than the British sample and far less trusting than Northern Irish Protestants. The attitudes of the two communities are far apart in this area. While almost half of Protestants think computers are not a serious threat or not at all a threat to privacy, only about a quarter of Catholics hold the same view. This provides a corrective to the figures showing low levels of trust that the British Government will act in the best interests of Northern Ireland. People in the sample clearly distinguish between the different arms and functions of government and while they may distrust government on the whole this can coexist with trust in, and loyalty to, particular arms of the government. Previous surveys have shown that on a range of questions measuring trust in government Northern Irish Protestants show higher levels of trust than people in GB and far higher levels than Northern Irish Catholics (Hayes and McAllister, 1996, p.19). The relatively high levels of trust by Protestants in this area probably reflects the fact that computer data and sophisticated intelligence information on individuals are perceived as key items in the arsenal of the state's campaign to preserve the existence of Northern Ireland, its 'war against terrorism', a campaign supported overwhelmingly by Protestants but not by Catholics.

Government and Dissent

Public attitudes to protest and to defy the law in varying degrees are often taken as providing a clear measure of more general attitudes to democracy and government. Opposition to public protest is often taken as a measure of authoritarian attitudes and of support for a strong repressive state. Attitudes to public protest also provide another measure of the level of trust in government. Support for a government's right to repress dissent implies a certain faith in the judgement of government and an acceptance of the authority of government. In the light of this, answers to questions on public protest in Northern Ireland are most interesting for the contradictions which they illuminate and for the questions they raise about the motivations which underlie these attitudes to protest. Responses to each question were first looked at according to religion and then contrasted with attitudes in GB. Responses are further explored according to attitudes to a united Ireland: taking attitudes to a united Ireland as an unsatisfactory but nonetheless useful surrogate for political identification.

There is a gap in attitudes between Catholics and Protestants in Northern Ireland, Catholics being notably more likely to agree that there are occasions on which people should follow their consciences even if it means breaking the law. Even so, Catholics are less likely to favour conscience over law than people in GB. The figures in Table 4.7 would suggest that Northern Ireland in general is a more authoritarian society than GB. When broken down according to attitudes to a united Ireland we get the somewhat surprising result that those most strongly in favour of a united Ireland (a group I will loosely call 'republicans') are a little less likely to favour conscience over law than those who merely support a united Ireland (a group I'll loosely call 'moderate nationalists') or those who don't care. This should remind us that the aspiration to a united Ireland is not necessarily a radical aspiration but can also spring from deeply conservative sentiments.

Table 4.7   **In general would you say that people should obey the law without exception, or are there exceptional occasions on which people should follow their consciences even if it means breaking the law?**

	NI %	NI Prot. %	NI Cath. %	NI other %	GB %
Obey the law	49	56	42	43	38
Follow conscience	51	44	58	58	62

... according to responses to the statement I would like the future of Northern Ireland to be within a united Ireland

	strongly agree %	agree %	neither %	disagree %	strongly disagree %
Obey the law	44	41	40	56	55
Follow conscience	56	59	60	44	45

In the light of the figures in Table 4.8 which show enthusiasm for public protest among those who strongly favour a united Ireland, the figures on 'conscience' suggest that favourable attitudes to protest in Northern Ireland should not be read automatically as evidence of anti-authoritarianism. Those most in favour of protest in Northern Ireland could easily become those most opposed to protest in another constitutional framework. When it comes to specific examples of public protest Northern Irish Catholics show themselves on balance a little less in favour of permitting protest than those in GB but often far more in favour than Northern Irish Protestants.

Table 4.8 shows that both Catholics and Protestants in Northern Ireland are somewhat less likely than people in GB to say public protest meetings should definitely be allowed. They suggest that the direct experience of civil unrest has contributed to hostile attitudes to public protest in Northern Irish and show only marginal differences in attitudes between Catholics and Protestants.

When analysed according to attitudes to a united Ireland however, certain gaps open up. Those strongly in favour of a united Ireland, the 'republicans', are the only group in Northern Ireland more in favour of allowing protest meetings than the British sample. As we move along the spectrum of attitudes to a united Ireland we see that the less people agree with a united Ireland the less likely they are to support protest, with a slight tilt up at the other end of the spectrum.

Table 4.8 **There are many ways people or organisations can protest against a government action they strongly oppose. Should the following be allowed ...**

Organising public meetings to protest against the Government

	NI %	NI Prot. %	NI Cath. %	NI other %	GB %
Definitely	48	48	47	53	57
Probably	39	37	44	33	33
Probably not	5	7	3	6	5
Definitely not	8	9	6	8	5

According to responses to the question I would like the future of Northern Ireland to be within a united Ireland

	strongly agree %	agree %	neither %	disagree %	strongly disagree %
Definitely	66	54	48	46	50
Probably	24	34	43	41	36
Probably not	2	6	4	5	6
Definitely not	8	6	5	8	8

Those most strongly opposed to a united Ireland, a group I will loosely call 'loyalists' are slightly more in favour of protest than those who simply oppose a united Ireland, a group I will loosely call 'moderate unionists'.

This pattern, of decreasing tolerance of protest as there is decreasing support for a united Ireland, with the tilt up at the end, remains consistent over attitudes to several different forms of protest. The relatively high support for protest among those most strongly supportive of a united Ireland and those most strongly opposed as compared to their 'moderate' fellows is best explained not by a general acceptance of protest in principle but by the experience of its recent application as a practical political tool.

Table 4.9   Allow . . .organising protest marches and demonstrations

	NI %	NI Prot. %	NI Cath. %	NI other %	GB %
Definitely	26	22	32	27	34
Probably	36	38	35	32	40
Probably not	17	16	16	20	13
Definitely not	21	23	17	21	12
Total not allow	36	39	33	41	25

According to responses to the statement 'I would like the future of Northern Ireland to be within a united Ireland'

	strongly agree %	agree %	neither %	disagree %	strongly disagree %
Definitely	52	34	30	15	23
Probably	21	36	35	38	42
Probably not	14	11	22	17	16
Definitely not	13	19	12	31	20
Total not allow	27	30	34	48	36

The differences which were visible in the question on meetings are more pronounced when marches and demonstrations are at issue. The total who say they would probably and definitely not allow the organising of protest marches and demonstrations is far higher in Northern Ireland than in GB. Catholics are more supportive than Protestants of the right to march and demonstrate, but not as supportive as people in GB. Only the strong supporters of a united Ireland are more likely to say they would definitely allow marches and demonstrations than are people in GB.

Analysed according to attitudes to a united Ireland we once again can see a pattern of decreasing support for protest according to decreasing support for a united Ireland with a tilt upwards at the end. However here the tilt at the 'loyalist' end of the scale is more dramatic. There are some massive gaps in attitudes, with over half of 'republicans' backing the right to protest 'definitely' compared to 15 per cent of 'moderate unionists'. Again the gap here is best explained not as a manifestation of a general authoritarianism or anti-authoritarianism but of approaches to very specific issues.

Table 4.10   **There are some people whose views are considered extreme by the majority. Consider people who want to overthrow the government by revolution. Do you think such people should be allowed to hold public meetings to express their views?**

	NI %	NI Prot %	NI Cath %	NI other %	GB %
Definitely	14	10	17	17	20
Probably	33	27	36	42	31
Probably not	21	23	23	9	17
Definitely not	33	40	23	31	32
Total not allow	54	63	47	40	49

According to responses to the statement 'I would like the future of Northern Ireland to be within a united Ireland'

	strongly agree %	agree %	neither %	disagree %	strongly disagree %
Definitely	28	24	10	10	10
Probably	37	43	41	28	24
Probably not	21	17	20	16	27
Definitely not	14	16	29	45	39
Total not allow	35	34	49	61	66

The final question examined in this area, and the first to explicitly mention revolutionaries, reveals even wider gaps in attitudes. While almost one half of the British sample would not allow revolutionaries to hold public meetings and about a third of 'republicans' and 'nationalists' feel the same, about two thirds of 'unionists' and 'loyalists' would not allow such meetings. There is no tilt at the end in this question; 'loyalists' are more likely than 'moderate unionists' to oppose this particular right. It is clear that the revolutionaries many of those in the Northern Ireland sample have in mind are republican paramilitary groups, as Gallagher has also suggested previously (1992, p.92).

Conclusion

Attitudes to government and politics in Northern Ireland exist in a context of fundamental uncertainty about the future form and framework of government. To both communities it appears that their repeatedly expressed democratic preferences on the fundamentals of government (which are of course at direct odds) have little political impact. It is easy for people to feel that their vote has little real impact. The low levels of interest in politics compared to GB may in part reflect the fact that people feel powerless to effect political change and see politics much more as something out of their control and government as something beyond their ability to influence. The fact that the conflict which began in 1969 has seemed to be more or less stalemated since the early 1970s has probably contributed to the low levels of interest and high levels of bewilderment.

The form of government which Northern Ireland has experienced for over a quarter of a century is a form of government rejected to one degree or another by every significant political force in Northern Ireland, including government itself. In the circumstances a lack of interest in and understanding of politics and high levels of distrust of government on the basic issue of whether government will act in the best interests of Northern Ireland are hardly surprising.

References

Breen, R. (1996), 'Who Wants a United Ireland? Constitutional Preferences among Catholics and Protestants' in R. Breen, P. Devine, and L. Dowds, (eds). *Social Attitudes in Northern Ireland : the Fifth Report, 1995 - 1996*, Blackstaff Press, Belfast.

Cochrane, F. (1997), *Unionist Politics and the Politics of Unionism since the Anglo-Irish Agreement*, Cork University Press, Cork.

Connolly, M. and Erridge, A. (1990), 'Central Government in Northern Ireland' in M. Connolly and S. Loughlin, (eds), *Public Policy in Northern Ireland: Adoption or Adaptation?*, Policy Research Institute, The Queen's University of Belfast and the University of Ulster, Belfast.

Gallagher, A.M. (1992), 'Civil Liberties and the State' in P. Stringer and G. Robinson, (eds), *Social Attitudes in Northern Ireland: the Second Report, 1991 - 1992*, Blackstaff Press, Belfast.

Hayes, B. and McAllister, I. (1996), 'Public support for democratic values in Northern Ireland' in R. Breen, P. Devine and L. Dowds, (eds), *Social Attitudes in Northern Ireland : the Fifth Report, 1995 - 1996*, Blackstaff Press, Belfast.

O'Leary, B. and McGarry, J. (1993), *The Politics of Antagonism: Understanding Northern Ireland*, Athlone Press, London.

Todd, J. (1990), 'The Conflict in Northern Ireland: Institutional and Constitutional Dimensions' in M. Hill and S. Barber, (eds), *Aspects of Irish Studies*, Institute of Irish Studies, Queen's University, Belfast.

Whyte, J. (1990), *Interpreting Northern Ireland*, Clarendon Press, Oxford.

5 Attitudes to the National Health Service in Northern Ireland

ANN MARIE GRAY

Background

In many respects Northern Ireland remains one of the most disadvantaged areas of the UK. The claimant unemployment rate is 11 per cent compared to 7 per cent in England, 8 per cent in Wales and 8 per cent in Scotland. Average gross weekly household income in Northern Ireland is £322.80 which is the lowest in the UK and over a fifth of the income of households in Northern Ireland comes from social security benefits - a higher proportion than for any other UK region (Regional Trends, 1997).

Health Profile of Northern Ireland

Evason and Woods (1995) note that despite its youthful demographic structure (26 per cent of the population consists of persons under the age of 16), on some indicators levels of health are lower than elsewhere in the UK. The infant mortality rate has improved significantly since the 1980s, when it was 13.2 per thousand live and still births, but it remains high in comparison to other regions of the UK at 7.6 per thousand live and still births (Regional Trends, 1997) which compares unfavourably with 6.2 in England and 6.2 in Scotland. The rate of perinatal deaths is 10.4 per thousand compared to 9.6 in Scotland and 8.8 in England.

Age adjusted mortality rates for 1995 (Regional Trends, 1997) show that Northern Ireland has high rates of mortality from Ischaemic Heart Disease, Cerebro Vascular Disease and for death by preventable causes such as road traffic accidents and poisonings. The Regional Strategy for Health and Social Wellbeing (DHSS, 1996) points out that nearly two thirds of all premature deaths in Northern Ireland are due to heart disease and cancers. The level of risk factors is high in the population - two thirds of the population aged 15-64 years have two or more risk factors for heart disease.

Analysis of the components of health and social need in Northern Ireland shows that in almost all areas the level of relative need is higher than in England. While relative need has fallen in relation to some components such as housing conditions and increasing income levels it has worsened in relation to other components such as the proportion of the population experiencing long term unemployment.

Structure of Health Services In Northern Ireland

The Department of Health and Social Services (Northern Ireland) is responsible for health, personal social services and social security. Arrangements for the provision of health and social services differ significantly from Britain. In Northern Ireland four area health and social services boards are responsible for the administration of both health and personal social services unlike the structure in Britain where responsibility for personal social services rests with local authorities. Since the restructuring of health care in the early 1990s and the introduction of the internal market, the health and social services boards have taken on a 'purchasing' role - they assess the needs of their resident populations and purchase health and social care services to meet those needs. In each board area health and social service councils have been established to represent the views of users. Hospital and community services are provided by twenty health service trusts - six community trusts, five integrated trusts, eight acute trusts and the ambulance trust. Just over 50 per cent of General Practitioners in Northern Ireland are fundholders (DHSS, unpublished) and 33 per cent of the population is covered by a fundholding GP (Regional Trends, 1997).

Eighty-five per cent of total health services expenditure in Northern Ireland is related to services provided by the NHS in the UK - the remaining 15 per cent covers personal social services. Acute hospital care absorbs 40 per cent of the total revenue budget for health and personal social services and Northern Ireland has one of the highest levels of acute hospital beds per head of population anywhere in the UK. In 1993 the rate of population to acute hospital bed was 270 to 1 in Northern Ireland; 400 to 1 in England. By 1996 the figure had increased to 322 to 1 in Northern Ireland, compared to 357 to 1 in the northern region of England (which has the same level of admission rates as in Northern Ireland) and 416 to 1 in England as a whole (DHSS, 1996).

This chapter uses the 1996 Northern Ireland Social Attitudes survey data to examine attitudes to health care provision. It assesses levels of satisfaction with the NHS and its component parts, it looks at attitudes to the funding of the service and whether or not the service should be restricted to lower income groups. Where possible, changes or trends in attitudes from previous surveys are charted and where appropriate, comparisons are made with the British data as reported by Judge, *et al* (1997).

Satisfaction with the NHS

Respondents were asked about how satisfied or dissatisfied they were with the way in which the NHS is run. Data collected in 1996 shows a clear drop in the overall level of satisfaction compared to 1991 and 1994.

Table 5.1 **All in all, how satisfied would you say you are with the way in which the NHS runs nowadays?**

	1991 %	1994 %	1996 %
quite/very satisfied	45	51	37
neither satisfied/dissatisfied	20	16	16
quite/very dissatisfied	35	33	47

This mirrors a similar downturn in support for the NHS in Britain with 50 per cent of respondents there expressing dissatisfaction - as in Northern Ireland the highest ever (Judge, *et al*, 1997). Prior to discussing possible reasons for this increasing dissatisfaction it is useful to look at attitudes to some of the component parts of the service. As in previous years people appear to be more satisfied with various aspects of the service than the service as a whole.

While decreasing slightly since 1991 and 1994, levels of satisfaction with GPs and NHS dentists remains at a consistently high level. In Britain the proportion of people expressing satisfaction with NHS dentists is considerably lower at 52 per cent. This highlights the steady decline which has taken place throughout the 1990s. Judge, *et al* (1997) indicate this may be due to the fact that as an increasing number of dentists withdraw from the

NHS, it is becoming increasingly difficult for new patients to obtain NHS treatment at all.

Table 5.2   Satisfaction with GPs and NHS Dentists

	1991 %	1994 %	1996 %
General Practitioners			
quite/very satisfied	85	85	83
neither satisfied/dissatisfied	6	6	5
quite/very dissatisfied	8	9	12
National Health Service Dentists			
quite/very satisfied	76	74	73
neither satisfied/dissatisfied	16	15	13
quite/very dissatisfied	8	11	14

Levels of satisfaction with hospital care are more in line with attitudes to the NHS as a whole. Campbell and Robinson (1993) noted that dissatisfaction with hospital services was particularly marked in the case of out-patient attendances.

Table 5.3   Satisfaction with in-patient and out-patient services

	1991 %	1994 %	1996 %
In-patient			
quite/very satisfied	73	74	65
neither satisfied/dissatisfied	15	13	16
quite/very dissatisfied	13	13	19
Out-patient			
quite/very satisfied	60	65	62
neither satisfied/dissatisfied	16	13	14
quite/very dissatisfied	25	22	24

In 1996 it remains the case that more people are dissatisfied with out-patient services than with in-patient services but the gap has narrowed, largely because of increasing dissatisfaction with in-patient care.

What particular aspects of the NHS are people most unhappy about? The quality of medical treatment and nursing care in hospitals does not come in for major criticism with 63 per cent of respondents stating that the quality of medical treatment in hospitals was 'satisfactory' or 'very good' and 72 per cent believing that this was the case for nursing care. On the other hand hospital waiting lists for non-emergency operations and staffing levels in hospitals attracted substantial criticism.

Table 5.4   % saying NHS needs 'a lot' or 'some' improvement

	%
Waiting time before getting appointments with hospital consultants	85
Hospital waiting lists for non-emergency operations	83
Staffing levels of nurses in hospital	75
Staffing levels of doctors in hospitals	74
Time spent waiting in Accident and Emergency departments before seeing doctor	74
Time spent waiting in out-patients departments	68
General condition hospital buildings	54
GPs appointment systems	46
Quality medical treatment in hospitals	37
Quality nursing care hospitals	28

In Northern Ireland there has been public concern about the cancellation of non-emergency surgery as Trusts have had to cancel or postpone operations. Statistics show that of the UK regions Northern Ireland has the highest percentage of people waiting more than six months for 'ordinary' admissions and day care admissions (Table 5.5).

Hospital Statistics for 1995 (DHSS, 1995) show a reduction of 23 per cent in the number of beds from 1990/91 with the total number of ordinary admissions increasing by 3 per cent and the number of day care admissions increasing by 15 per cent. The statistics also point to increased throughput which the DHSS states has been made possible by reducing both length of stay and the period a bed remains unoccupied between patients. The average number of ordinary admissions treated in each available bed increased by 47 per cent since 1990/91 and 8 per cent between 1993/94 and 1994/95. Between 1990/91 and 1994/95 the average length of stay for all specialities decreased by 33 per cent. It may be the case that reduced stay is unpopular and raises concerns about patients being discharged before they are ready.

Table 5.5   Waiting times for hospital admissions

	waiting 6 months or more %	waiting 12 months or more %
Ordinary admissions		
England	24.6	1.9
Wales	24.6	11.6
Scotland	16.7	1.4
Northern Ireland	25.9	12.6
Day care admissions		
England	17.8	0.8
Wales	17.8	5.8
Scotland	9.9	0.7
Northern Ireland	20.2	8.6

Source: Regional Trends, 1997

Do factors such as age, gender and social class have any bearing on levels of satisfaction? Campbell and Robinson (1993) commented that satisfaction with the NHS increased with age (their age bands were slightly different) but this would no longer appear to be the case. Table 5.6 shows that satisfaction with the NHS decreases between 30 and 69 years of age rising again for those aged 70 and over.

Table 5.6   Satisfaction with the NHS by age group

	18-29 %	30-49 %	50-69 %	70+ %
quite/very satisfied	44	33	30	56
quite/very dissatisfied	41	51	52	34

The over 70s are likely to be the heaviest users of the service and they may appreciate the benefits of the NHS. The age group which appears most dissatisfied were the 50-69 age group.

Those in social classes I and II are more likely to express dissatisfaction. This could be due to such people having higher expectations and /or perhaps being more aware of ideological arguments about the NHS and the funding of the service. An additional factor could be that the privatisation of parts of the service such as dentistry and opthalmics may have impacted more on these groups.

Table 5.7   Satisfaction with the NHS as a whole and with particular aspects of the service by social class

	I	II	III nm	III m	IV	V
	%	%	%	%	%	%
National Health Service Overall						
quite/very satisfied	19	28	41	42	41	36
quite/very dissatisfied	68	59	45	42	38	47
General Practitioners						
quite/very satisfied	65	81	88	83	88	73
quite/very dissatisfied	16	12	11	13	6	22
National Health Service Dentists						
quite/very satisfied	61	66	74	77	75	73
quite/very dissatisfied	29	18	13	9	17	6

Levels of satisfaction were also considered in relation to unionist/ nationalist affiliation. A fairly high proportion (44 per cent) of respondents described themselves as neither nationalist or unionist. Of those who described themselves as nationalist or unionist a higher proportion of respondents in both groups were more dissatisfied than satisfied with the NHS as a whole. Both groups expressed high levels of satisfaction with GPs although the figure was slightly higher for unionists - 85 per cent as compared to 78 per cent for nationalists. Seventy three per cent of both groups were satisfied with NHS dentists.

What would explain the increasing dissatisfaction with the NHS as a whole but continuing high levels of satisfaction for many of its component parts? Obviously the concern about hospital waiting lists discussed earlier is one factor but it is possible that people were dissatisfied with aspects of the service they were not directly asked about. This could be particularly relevant in Northern Ireland where, due to the integrated structure, services such as home helps, community care and residential and nursing home provision may be perceived as health care rather than social care issues. The financing of residential care for elderly people received considerable media attention in 1996 when in response to increasing public concern government published a policy statement 'A New Partnership for Care in Old Age' (HMSO, 1996). This emphasised the importance of individual savings, pensions and care insurance to security in old age. This was perceived by some as government withdrawing from its obligations and failing to fully recognise and give regard to the contributions made during a lifetime.

The picture generally is quite different from findings of the 1991 and 1994 surveys (Campbell and Robinson, 1993; Largey and O'Neill, 1996) when satisfaction appeared to remain fairly stable and indeed increase in some cases between 1991 and 1994. It is possible that increased levels of dissatisfaction could be due to changes which had occurred or were planned or implemented during the year the data was collected. For example, in 1996 the Eastern Health and Social Services Board published a consultation paper on the reorganisation of acute hospitals in Belfast (EHSSB, 1996). A number of the proposals contained in the document generated controversy and received considerable media attention, in particular those concerning the proposed relocation of paediatric and maternity services. Also in 1996 the DHSS published its regional strategy for health and social well being covering the period 1997-2002 (DHSS, 1996) which emphasised the need to reduce acute hospital provision in Northern Ireland and centre acute care around the core framework of the two major teaching hospitals in the province - the Royal Group of Hospitals Trust and the Belfast City Hospital Trust, plus four other acute hospitals. As the overall aim was the reduction in acute services this proposal was unpopular in areas where provision would cease to exist or would be restricted. Public opposition to hospital planning was expressed through campaigns such as the action group set up to prevent the closure of Banbridge Hospital or to ensure the building of a new hospital at Downpatrick.

Did the health care reforms introduced during the early 1990s result in reduced levels of satisfaction? It is not possible to say for sure as there has not as yet been a comprehensive assessment of the reforms in Northern Ireland. However, research by Mulholland and McAlister (1997) concludes that the benefits which should have come out of the restructuring are yet to be fully realised. At the heart of the restructuring of health care was the assumption that moves to a market cen..red system would lead to improved service efficiency and responsiveness via the separation of purchaser and provider.   Mulholland and McAlister (1997) assess the extent to which the internal market for acute services in Northern Ireland has met the expected outcome of efficiency, responsiveness, choice, quality and equity.   The providers, purchasers and GP fundholders who participated in the study felt that the internal market had improved responsiveness but the results relating to choice were not very positive.   It would seem from this research in Northern Ireland and from research in Britain (LeGrand and Robinson, 1994; Ranade, 1994) that there are a number of obstacles in theory and practice to the achievement of the objectives set out in the NHS reforms.

Attitudes to Public Spending and the NHS

As levels of satisfaction have decreased the proportion of people who see health as a priority for public spending has increased.   As in previous years respondents have made spending on health care the number one priority for extra spending.

Table 5.8   Respondents' choices for first priority for extra government spending*

	1991	1994	1996
	%	%	%
Health	49	56	63
Education	21		21
Housing	7		4
Social Security Benefits	10		4

*   Figures not available for education, housing and social security for 1994

Since 1991 there has also been an increase in the number of respondents placing health as a first or second priority.   In 1991 the figure was 73 per cent and this has risen to 88 per cent in 1996.   This may suggest that respondents associate decreased levels of satisfaction with a problem of inadequate funding.   It could also be due to increased media attention about the cancellation of elective surgery and publicity about the financial crisis in a number of Trusts.   Regardless of age, class, political affiliation or gender, health care was accorded the highest priority for extra government spending.

This demonstrates a degree of consensus which does not apply to the other areas of public spending respondents were asked about. For example, a fairly strong class difference can be detected in relation to attitudes to housing and social security spending. This is likely to be due in part to the fact that housing and social security are selective services and benefits available only on a test of means - unlike the universality of the NHS.

Despite decreasing satisfaction respondents obviously still strongly favour the provision of health care on a universal basis. There was considerable opposition to the suggestion that the NHS should in future be available only on lower income groups. This opposition also applied to dental services. Similar attitudes were expressed in Northern Ireland and GB.

Table 5.9   NHS should be available only to those on lower incomes

	NI	GB
	%	%
Support	25	21
Oppose	72	77

It is pertinent to note that while 72 per cent of Northern Ireland respondents opposed this statement, 56 per cent actually opposed it a lot which is some indication of the strength of feeling which exists on the subject.

Patient Choice

A stated aim of the restructuring of the NHS in 1989 was to improve patient choice. The Patients Charter for Northern Ireland was published in 1992. It signalled the government's commitment to: improve the quality of all public services in Northern Ireland; make services more responsive to the needs of individual citizens and ensure value for money. In the light of this it is interesting to gauge respondents opinions on the degree of control or choice they feel they have gained. The questions in this survey do not allow for a comprehensive analysis of this issue but two questions give some indication on attitudes to choice in relation to GP and hospital care. Sixty eight per cent of respondents believed they could change their GP without too much difficulty but only 40 per cent believed they would have a say in which hospital they were sent to. This suggests that people do feel they have more choice in relation to primary care. The belief that patients have more restricted choice in relation to hospital care would seem to be supported by research on the restructuring of the health service and the working of the internal market (LeGrande and Robinson, 1994) which found that it is GP fundholders who have the choice of deciding where patients receive care, not the patient.

Of the many concepts and issues which have dominated health care debates in developed welfare states few have received as much attention as patient choice and consumer participation. A number of commentators have argued that patient involvement in health care evaluation is usually confined to being the occasional respondent to surveys about health services or being part of the consumer-oriented developments in the NHS which have top-down paternalistic origins (Williams, 1994; Klein, 1995). However, as Fitzpatrick and White (1997) point out, the argument that patients have been confined to a narrow, passive and occasional role in the evaluation of their services presupposes that patients would readily play a more active role in this regard. Some studies such as that carried out by Richardson, *et al* (1992) showed that a majority of the general public surveyed as part of their research wanted greater public participation about services generally. On the other hand a number of studies show that patients do not act like consumers and for example, do not seek decision-making active choices about doctors (Salisbury, 1989; Abelson, *et al,* 1995).

The Rationing and Prioritising of Health Care Resources

Research in the UK has generally shown that people expect a comprehensive publicly funded health service and are against the restriction of non-essential treatments. There has also been little support for eliminating any treatment from the NHS (Davies, 1991; Salter, 1992). Rationing - mainly in the form of waiting lists has always existed in the NHS but the debate about rationing has gained considerable force in recent years.

The restrictions on treatment have become more obvious as the gap between available resources, needs and expectations have widened. Mohan (1995) has argued that additional rationing devices have been introduced. First, the restructuring of the health service has resulted in rationing decisions being made as a result of specific strategies - for example by explicitly defining how much of what service should be provided through the contract system.

Table 5.10   The prioritising of health care

	%
Age	
Who would get priority?	
The younger man	36
the older man	7
age would make no difference	47
can't choose	10
Who should get priority?	
The younger man	20
the older man	6
age should make no difference	64
can't choose	10
Weight and Diet	
Who would get priority?	
Average weight, eats healthily	45
overweight, eats unhealthily	3
weight would make no difference	41
can't choose	11
Who should get priority?	
Average weight, eats healthily	29
overweight, eats unhealthily	2
weight should make no difference	57
can't choose	11
Smoking	
Who would get priority?	
Non-smoker	54
heavy smoker	3
would make no difference	33
can't choose	11
Who should get priority?	
Non-smoker	38
heavy smoker	2
should make no difference	51
can't choose	9

Second, some services have been abandoned because they are deemed 'cosmetic' and third there has been some discussion of rationing services according to age, or refusing treatment to people whose Second, some services have been abandoned because lifestyle seems likely to prejudice their ability to benefit from it, such as refusing smokers coronary bypass surgery.

What do people in Northern Ireland think about the prioritising of health care? Survey respondents were asked for their opinions in relation to three situations (Table 5.10).   Each situation involves two men with heart conditions. Both were awaiting surgery and both would benefit from the operation. In the first situation one man is aged 40 and the other 60; in the second situation one man is of average weight and eats healthily but the other is very overweight and eats unhealthily; in the third situation one man smokes heavily and the other is a non-smoker. People were asked whether or not the differences should be used to prioritise between the two men in terms of the operation. As Table 5.10 shows people are not supportive of the idea of prioritising on the basis of lifestyle or age in relation to the situations set out. People believe this form of rationing happens to a greater extent than they believe it should. There is more support for giving priority to non-smokers although the majority of respondents still feel this should not be the case.   This could be seen as a further endorsement of support for the universality of the service.

Conclusion

The NHS has been in an almost constant state of review and change since the 1980s. Levels of dissatisfaction clearly rose between 1994 and 1996 and this was accompanied by increasing support for greater funding to the service. People remain strongly in favour of a comprehensive and universal system of health care.   Judge, et al (1997) suggest that it would not be surprising if dissatisfaction with health care was one of the factors which contributed to the defeat of the Conservatives at the 1997 General Election. At the time of writing (December, 1997) the NHS is about to be subjected to more changes.   In December 1997 the Labour Government published a White Paper on the NHS which in government's words will abolish the internal market in health care. A set of structural changes are due to be implemented in April 1999. Under these the commissioning of health care will be taken over by primary care groups led by GPs and community nurses and this will replace the current system of commissioning by GP fundholders and health authorities. The separation of the planning and provision of health care will remain although the rhetoric now centres on 'partnership' and 'collaboration' rather than competition. Also included in the White Paper is the setting up of two new national bodies tasked with the responsibility of achieving improvements in the cost-effectiveness and quality of care. These proposals have received a cautious welcome from health care professionals and policy analysts but it will remain to be seen if these changes result in a positive shift in public attitudes.

References

Abelson, J., Lomas, J., Eyles, J., Birch, S. and Veenstra, G. (1995), 'Does the Community Want Devolved Authority?', *Canadian Medical Association Journal*, vol.153, pp. 403-412.

Campbell, R. and Robinson, G. (1993), 'Who Cares for the National Health Service?', in P. Stringer, and G. Robinson, (eds), *Social Attitudes in Northern Ireland: the Third Report*, Blackstaff, Belfast.

Davies, P. (1991), 'Thumbs down for Oregon Rations', *Health Service Journal*, 14 Nov.

Department Health Social Services. (1995), *Hospital Statistics 1993-1995*, DHSS, Belfast.

Department Health Social Services. (1996), *Health and Wellbeing into the Next Millenium*, DHSS, Belfast.

Eastern Health Social Services Board. (1996), *Standing Together*, EHSSB, Belfast.

Evason, E. and Woods, R. (1995), *Poverty, Charity and Doing the Double*, Avebury, Aldershot.

Fitzpatrick, R. and White, D. (1997), 'Public Participation in the Evaluation of Health Care', *Health and Social Care in the Community*, vol.5, no.1, pp. 3-8.

HMSO (1996), *A New Partnership for Care in Old Age*, HMSO, London.

Judge, K., Mulligan, J.A. and New, B. (1997), 'The NHS: New Prescriptions Needed?', in R. Jowell, J. Curtice, A. Park, L, Brook, K. Thomson, and C. Bryson, (eds), *British Social Attitudes: the 14th Report* : Ashgate, Aldershot.

Klein, R. (1995), *The New Politics of the NHS*, Longman, London.

Largey, A. and O'Neill, C. (1996), 'Satisfaction with Health Services in Northern Ireland', in R. Breen, P. Devine, and L. Dowds, (eds), *Social Attitudes in Northern Ireland: the Fifth Report*, Appletree Press, Belfast.

LeGrand, J. and Robinson, R. (eds), (1994), *Evaluating the NHS Reforms*, Kings Fund Institute, London.

Mohan, J. (1995), *A National Health Service? The Restructuring of Health Care in Britain since 1979*, Macmillan, Basingstoke.

Mulholland, G. and McAlister, D. (1997), 'The Quasi-market in Health Care: Pre-requisites, Problems and Prospects', *Papers in Public Policy and Management*, no.66, University of Ulster, Jordanstown.

Ranade, W. (1994), *A Future for the NHS: Health Care in the 1990*, Longman, New York.

Richardson, A., Charny, M. and Hammer-Lloyd, S. (1992), 'Public Opinion and Purchasing', *British Medical Journal*, vol.304, pp. 680-682.

Salisbury, C. (1989), 'How do people choose their Doctor?', *British Medical Journal*, vol.299, pp. 608-610.

Salter, B. (1992), 'Heart of the Matter', *Health Service Journal*, 1 Oct.

Williams, B. (1994), 'Patient Satisfaction: A Valid Concept?', *Social Science and Medicine*, vol.38, pp. 509-516.

6 Attitudes to the Environment in Northern Ireland

ADRIAN MOORE, SALLY COOK AND CLAIRE GUYER

Introduction

The past two decades have seen a substantial growth in public awareness of environmental issues, evolving essentially from fashionable social ideals into much more important pragmatic, political and economic concerns. It is, however, well recognised (Smith, 1993; McGrew, 1993; Lowe and Goyder, 1983) that concern for the environment is not a recent phenomenon and that political activity focusing on environmental issues and problems has its roots within the Industrial Revolution. Until recently, discussion of this concern commonly focused on its cyclical nature, with apparent periods of heightened environmental concern appearing to coincide with economic prosperity. McGrew (1993) argues that there is now strong evidence to suggest that we are moving out of this cyclical pattern into a new era of concern for the environment such that it is likely to be a permanent feature of our society and of the political and policy-making agendas. The reasons cited for this apparent transformation in attitude relate to the increasing complexity, scale and severity of environmental problems and the gradual recognition at the global level that there is an undeniable inter-linkage between economic, social and environmental issues. This recognition culminated in the United Nations Conference on Environment and Development (UNCED, commonly known as the Earth Summit) held in Rio de Janeiro in 1992 and in the production of the Agenda 21 agreement. It is this document, and the subsequent monitoring of progress since UNCED (Department of the Environment, 1996), that has created the legislative and policy-making context within which governments are currently working.

Within the UK much of the impetus for environmental policy-making now rests with the European Union (EU). The Fifth Environmental Action Plan and its commitment to the integration of environmental concerns are driving current environmental policy across previously independent policy sectors. Policy decisions taken at EU level and translated into EU directives and regulations are subsequently enacted within the UK by incorporation into state legislation, and eventually reach Northern Ireland through an Order in Council. In addition, as a response to UNCED and Agenda 21, in 1994 the UK government published its strategy for moving towards sustainable development which included policy required within Northern Ireland (HMSO, 1994).

Within Northern Ireland the legislative and policy framework for the environment is conditioned extensively by both the GB and EU contexts. There is usually a significant time lag between Acts being debated and agreed in the Westminster Parliament and the corresponding Order in Council being written for Northern Ireland. Whilst this may have disadvantages for environmental managers in Northern Ireland, it may also provide the opportunity to adapt GB legislation into a more appropriate form for the province.

It is recognised that Northern Ireland does not experience, to the same form or extent, many of the environmental problems associated with modern industrialisation and urbanisation seen in GB. Recognition of the environment as an important economic resource, promotion of the 'clean environment' as a key marketing strategy and the potential to expand the 'green economy' (Department of Economic Development, 1993) elevate environmental issues to a more prominent place on the province's political agenda. A major concern therefore for the development of Northern Ireland's economic future is a requirement for effective environmental protection measures to maintain current favourable conditions and prevent serious environmental problems arising.

Continued popularity of environmental issues on the political scene depends to a large extent on the continuing strength and airing of public attitudes. In fact, governments wishing to implement more stringent and costly environmental legislation need to do more to increase public understanding of the importance of protecting the environment to make the policies actually work in practice (Witherspoon, 1995). Surveys such as the International Social Survey Programme, Eurobarometer and the British and Northern Ireland Social Attitude surveys play a significant role in establishing the public mood and in monitoring change over time. This pivotal role of continued public interest and support of environmental matters and the formal introduction of environmental issues via compulsory legislation into national and local political agendas form the background to this chapter.

About this Chapter

This chapter incorporates a predominantly descriptive analysis of a selection of general environmental issues covered in the 1996 Northern Ireland Social Attitudes survey. Primarily, attention will focus on attitudes towards the general theme of responsibility for environmental protection and in particular, on opinions of the respective roles played by government, business and industry, independent environmental groups and the individual. The availability of attitudinal data from previous years, namely: Stringer (1992); Yearley (1995); and Bryson, et al (1997), in addition to comparable data for GB, affords the opportunity to conduct some basic spatial and temporal analysis. Such analysis is useful in that GB can be used as a benchmark against which changing attitudes in the province can be examined. Of interest will be the way in which, if at all, the attitudes of Northern Irish people have changed both over time and relative to those in GB. Stringer (1992) proposed that the Northern Irish public were much less concerned about environmental issues than the British and suspected that a trickle-down effect from the mainland permeating social and geographical barriers would take effect over time, resulting in parity of awareness, understanding and attitudes. Indeed, Yearley's (1995) examination of the 1993 survey data provides some evidence to support this theory.

General attitudes, temporal trends and a comparison with GB form the main thrust of the investigation, although some basic sub-division and two-way crosstabulations by selected socio-economic variables will be incorporated. The very factual and descriptive nature of the analysis has been fashioned by the decision to concentrate on an examination of the complex array of temporal and regional comparisons. Certain variables though will be examined on selected statements (mainly for Northern Ireland) to identify those factors with the most discriminating influence on people's attitudes. Key categories include socio-economic status, educational qualifications, age, gender, religion and place of residence, all of which have been identified as significant in previous surveys. It is worth noting that certain socio-economic factors may help explain observed differences in attitudes within Northern Ireland and in comparison with GB. For example, Northern Ireland has a higher proportion of its population in the lower socio-economic classes and in the younger age groups (18-30) than GB. Fewer people in Northern Ireland (approximately 5 per cent fewer) have obtained educational qualifications at or above third level standard and a substantially larger proportion (8 per cent more) have no formal qualifications at all. While there are slightly fewer people proportionately living in city and suburban environments, 16 per cent more people (20 per cent in total) live in the countryside.

The Role of Government and the Individual in Environmental Responsibility

In this section Northern Irish attitudes on the broad environmental issues of responsibility and action concerning the government and the individual will be examined and compared with those in GB. This will serve a useful purpose in providing an initial impression of people's attitudes. Also, the fact that some of the questions were asked in previous surveys facilitates, at the very least, a cursory examination of temporal change in attitudes.

(i) Government

Table 6.1 presents the responses for Northern Ireland and GB to the statements regarding government spending on the environment and its responsibility to control environmental damage caused by industry.

There is very little difference between Northern Ireland and GB in overall opinions (around 90 per cent in the combined 'probably' and 'definitely' categories) regarding the responsibility of government to reduce levels of industrial damage to the environment. There is however a very noticeable difference in attitudes on the issue of government spending. The Northern Irish respondents are more inclined to believe that government should spend more or much more on the environment than their British counterparts. Comparison with the 1990 survey reveals that, while there has been a significant overall reduction in support for the idea of increased government spending in both countries, the nature of the change between the two is most interesting. The figure for Northern Ireland in 1990 was 54 per cent, and for GB 61 per cent (Stringer, 1992, p.36), meaning there has been a reduction of 7 per cent and 20 per cent respectively in support of increased spending over the intervening period. The very dramatic decline in the British figures has meant that those in Northern Ireland are now relatively more supportive of spending by government than the British.

Table 6.1   (a) **Would like to see more or less government spending on the environment**

	NI	GB
	%	%
Much more	10	8
More	37	33
Same as now	39	47
Less	5	4
Much less	1	1

(b) **Government responsibility to impose strict laws to make industry do less damage to the environment**

	NI	GB
	%	%
Definitely should be	62	58
Probably should be	28	31
Probably should not be	4	4
Definitely should not be	1	1

It is quite possible that the British are more inclined to believe that the level of government spending on the environment at the time was adequate and that they fear more spending would lead to higher taxes which were already viewed as being too high. In such circumstances, other key concerns such as health, education and social services are also competing for limited government funding. It is well recognised (Christie, 1996) that such concerns usually take priority over the environment.

Examining possible associations with social and economic factors in the Northern Ireland survey, two in particular emerge as the most significant, namely social class (Registrar General's classification divided into four categories) and educational attainment (Table 6.2). There is an obvious social class trend whereby the higher classes are substantially more in favour of increased government spending with classes I and II at least 10 per cent more in favour of an increase than the rest. A very similar trend is apparent in the education level data with those attaining qualifications from 'A' level upwards, being much more in favour of increased spending. Most noteworthy is the extent of support for much more spending (20 per cent) for those educated to at least degree level.

Table 6.2   Would like to see more or less government spending on the environment by social class and educational level

	much more		more	same	less	much less
	%		%	%	%	%
Social Class						
I + II	10	(55)	44	37	3	1
III NM	5	(45)	40	42	2	1
III M	15	(40)	26	41	3	2
IV + V	8	(41)	33	40	9	1
Education						
Degree	20	(69)	49	29	-	-
3rd Level	4	(49)	44	47	1	-
'A' Level	8	(54)	47	31	2	1
'O' Level	4	(40)	37	47	4	1
None	14	(43)	29	37	8	2

Note: the figures in brackets are the results of analysis using the combined 'more' and 'much more' categories and may contain different proportions to the simple addition of individual category proportions on account of the weighting of the data.

These figures in themselves are what we might have logically expected. In comparison with the British data, however, a few unusual facts emerge. Given that within Northern Ireland, there are proportionately fewer people in the higher social classes and fewer with degree and third level qualifications, it might be expected that attitudes would be less favourable towards increased spending than those in GB. As observed in Table 6.1, quite the opposite is true.

The differences in strength of opinions in GB are observable through all the class ranges and education categories. Support for an increase in spending was 46 per cent for social classes I and II and 41 per cent among classes IV and V. Breaking down the data for education the figures were 52 per cent for degree level and 35 per cent for those with no qualifications. The differentials between the highest and lowest social classes and education categories for the two countries are interesting. For social class, the differentials are 13 per cent in Northern Ireland and 5 per cent in GB and for education they are 26 per cent and 17 per cent respectively. The conclusion is that class and education tend to have a greater influence on attitudes to government spending in Northern Ireland than they do in GB. One reason may be a consequence of environmental issues having a lower overall profile in Northern Ireland and, given that environmental degradation is less obvious, it is therefore not unreasonable to expect that the best educated would be better informed compared to the majority of the population. Furthermore, the tax factor may have a much more significant and pervasive impact on the British public than on those in Northern Ireland.

Table 6.3 **Increase or decrease in government spending on the environment by place of residence**

	much more		more	same	less	much less
	%		%	%	%	%
Big city	11	(40)	29	34	6	-
Suburbs	6	(46)	41	41	4	-
Town	12	(52)	41	38	3	1
Village	13	(61)	48	21	2	-
Country	8	(30)	21	52	9	2

Place of residence (Table 6.3) was another interesting variable, which was peculiar to Northern Ireland in having an influence on opinion. Two points are worth mentioning. Firstly, there appears to be a very distinctive urban to rural trend in attitudes with opinions towards increased spending generally gaining favour the less urban the setting. The second point refers to the very obvious exception to this observation whereby those living in the most rural areas have the lowest level of support for increased spending, with a majority in favour of the status quo and the highest proportion of all groups (11 per cent) opting for lower spending. The first point is relatively easy to explain given the now widely accepted fact that those living in urban and metropolitan areas are in general less involved in, and possibly less concerned with environmental matters (social class and educational attainment factors probably account for this). The second point remains somewhat elusive in terms of a logical explanation. A number of options were investigated including the most obvious possibility that those engaged in farming may hold distinctly separate views. Indeed, after removing farmers from the analysis, the difference between country dwellers and the rest actually increased. The relatively high proportion of the higher socio-economic classes and the better qualified living in the countryside compared to most other categories confuses the issue even more.

On the other question regarding whether or not government should be responsible for making industry do less damage to the environment, similar trends were found, with social class, education and place of residence all having an influence on opinions but to a much lesser degree. In all three cases, there were approximately 10 per cent differences between the extreme categories. Overall, support in favour of government being responsible was very high, ranging from 95 per cent to 85 per cent across the social classes, 90 per cent to 80 per cent between places of residence and 72 per cent to 62 per cent from most to least qualified.

All told then, while an overwhelming majority of the population in both Northern Ireland and GB believe it is the government's responsibility to control the effects of industry on the environment, a much smaller proportion in both countries believe that the government should be spending more on environmental matters. Social status, education and area of residence play more significant roles in influencing opinion in Northern Ireland than in GB. The convergence of opinions between the two countries regarding levels of government spending is seen as a consequence of a more remarkable swing in public opinion in GB since 1990 than in Northern Ireland.

(ii) The Individual

The strength of opinions expressed by the Northern Irish respondents towards government might be better understood with reference to Table 6.4 which demonstrates relatively lower levels of confidence in the ability of the individual to do much for the environment compared to GB. Of the Northern Irish sample, 34 per cent agreed and 9 per cent strongly agreed that it was too difficult for the individual to do much for the environment compared to 27 per cent and 7 per cent in GB, while around 30 per cent either disagreed or strongly disagreed compared to almost 40 per cent in GB. These figures appear somewhat ironic given that a slight majority of the Northern Irish sample (54 per cent in all) claim to do what is right for the environment, even if it costs. The impression given is that, despite a significant proportion acting in the best interests of the environment, there is less confidence among the Northern Irish in the ability of the individual to make a difference. This can be qualified with the popular feeling, identified above, that the government should do more in terms of financial investment in environmental matters. It reinforces the notion that individuals within

Northern Ireland feel they have a limited role to play in government as so many decisions are already taken at national level and the fact that fewer powers are devolved to local authorities. Consequently, people have little choice but to continue to rely on and trust regional government to make decisions on their behalf.

Table 6.4   (a) It is just too difficult for someone to do much about the environment

	NI %	GB %
Strongly agree	9	7
Agree	34	27
Neither	21	20
Disagree	25	30
Strongly disagree	5	8

(b) I do what is right for the environment, even when it costs more money or takes more time

	NI %	GB %
Strongly agree	8	7
Agree	45	44
Neither	28	30
Disagree	8	8
Strongly disagree	1	1

In comparison with the 1995 Northern Ireland survey (Bryson, *et al*, 1997) there has been a small but noticeable swing towards the middle ground on both statements with a 4 per cent increase in the 'neither' category. A 3 per cent drop in those disagreeing that it is too difficult for the individual to do much for the environment was the largest change in response. It does appear then that there is a slight move towards reduced confidence among the population of the province in the ability of the individual to do much for the environment. Countering this to some extent though, is the substantial proportion of people who continue to claim to be environmentally conscientious in their daily lives.

Similar to the statements on government, very clear trends in attitude towards the individual's impact on the environment are apparent when the data is broken down by both social class and education (See Table 6.5). The higher social classes (I and II) are much less inclined to believe that it is too difficult to do much for the environment than the lower class groupings. Classes III manual, IV and V were at least 20 per cent more likely to agree with the statement while those in III non-manual were the most likely at 30 per cent, to neither agree or disagree. Levels of educational attainment have the most extreme variation with over 50 per cent of those educated to degree level disagreeing or strongly disagreeing with the statement, compared to less than 20 per cent of those with no formal qualifications. A general pattern is evident with the noted exception of those qualified to third level standard who appear to be less confident of their ability to do much for the environment than those educated to 'A' level standard. Age displays a similar pattern with a very pronounced rise in levels of agreement with increasing years. The older the respondent, the greater the likelihood of their believing that it is too difficult to do much for the environment.

Comparing the British and Northern Irish surveys, the lower social classes in GB were less likely to agree with the statement (42 per cent compared to 55 per cent in Northern Ireland in classes IV and V) while the difference between the highest classes (I and II) was not as strong (25 per cent compared to 31 per cent in Northern Ireland). Educational attainment displayed fairly similar trends, although those with no formal qualifications in GB were more likely to disagree with the statement. Similarly, while age was a less influential factor in the British data, the two oldest age groups were much more likely than the others (between 10 – 15 per cent) to disagree with the statement. The conclusion therefore is that in Northern Ireland, older age, low levels of educational qualifications and low socio-economic status are more significant factors in determining negative or sceptical attitudes towards the individual and what they can do for the environment than in GB.

Indeed, these figures lend some support to Yearley's (1995) suggestion that the best explanation of differing attitudes between Northern Ireland and GB can be found in the differences between the respective manual class groupings. This trend though is less clearly apparent in other analyses in this chapter.

Table 6.5 **It is just too difficult for someone to do much about the environment by social class, educational level and age group**

	strongly agree		agree	neither	disagree	strongly disagree
	%		%	%	%	%
Social Class						
I + II	5	(31)	26	21	35	9
III NM	2	(35)	32	30	28	4
III M	22	(58)	37	14	19	3
IV + V	10	(55)	46	18	15	3
Education						
Degree	2	(17)	15	30	36	15
3rd Level	4	(39)	35	25	28	3
'A' Level	4	(26)	21	24	44	4
'O' Level	9	(44)	34	23	25	4
None	13	(56)	43	16	15	4
Age Group						
18-30 yrs	8	(31)	23	30	30	8
31-45 yrs	6	(34)	28	23	30	7
46-60 yrs	7	(49)	43	19	22	3
61-90 yrs	17	(66)	49	10	17	1

With respect to the statement on whether the individual respondents do what is right for the environment despite the cost, similar but less clear patterns emerge. The two highest social classes (I and II) and the two most highly educated groups (degree and third level) had the greater proportions claiming to do right for the sake of the environment. Interestingly, in respect of age, it was the 31 to 45 year olds who, with 58 per cent, had the highest proportion making the same claim. It might well be that this age group would be among those who, being conscious of the importance of the environment, are in a better position financially to bear the added costs of taking the more environmentally friendly action. When other factors were considered, place of residence showed a noticeable distinction with those living in the city and suburbs being approximately 13 per cent less likely than the rest to agree. Females were 8 per cent more likely to do the right thing and while there was no noticeable difference between Catholics and Protestants, those respondents categorised as 'other' in religious terms were substantially more likely to claim to act in the interest of the environment.

Overall, these results give some cause for optimism for the future given the encouraging responses of the young and the better educated and the fact that a majority of people claim to act in the interests of the environment even at the risk of higher personal costs. A slight cause for concern though might be the relatively low levels of belief in the ability of the individual to do much for the environment and the early indication that environmental support may be falling.

Trust and Environmental Decision-making

Further insight can be gained when we examine the responses to questions on who the public trust to make the right decisions about the environment, and to what extent. Table 6.6 presents levels of trust in both Northern Ireland and GB for the years 1993 and 1996 with respect to scientists, business and industry, environmental groups, the government and people. Responses to the latter two groupings lend further support to earlier findings. In Northern Ireland, survey respondents are much more inclined to have some trust in government compared to GB while attitudes towards ordinary people are virtually identical in all respects. Attitudes over the three-year period remained remarkably consistent on these questions with the exception of a slight trend towards more trust in people in the British survey. The extremely large gap in overall levels of trust in government between Northern Ireland and GB remains and, if anything, is growing slightly (over 50 per cent have some or a lot of trust in government in Northern Ireland compared to less than 40 per cent in GB). The relatively high level of trust in Northern Ireland is somewhat unusual given that in other public opinion polls, politicians were consistently revealed as being the least credible group when it comes to decision-making (O'Riordan, 1995).

Further examination of the Northern Irish data support some of Yearley's findings, from the 1993 survey. A similar social class effect was evidenced with the higher social classes being more trusting, although social class differences were less pronounced than expected. Protestants were also found to be much more trusting of government (67 per cent) than the other two groups, Catholics in particular (40 per cent) having the least trust (14 per cent having no trust at all). This certainly suggests that religio-ethnic divisions of trust in government in Northern Ireland cut across more than political and constitutional issues and may go some way to explaining the observed differences with the British survey.

Table 6.6   **Level of trust to make the right decisions about the environment**

	a lot		some		very little		none	
	1993	1996	1993	1996	1993	1996	1993	1996
	%	%	%	%	%	%	%	%
Scientists								
NI	11	15	58	54	16	14	4	6
GB	14	18	57	59	18	11	2	3
Business and industry								
NI	2	1	35	27	41	46	13	17
GB	1	2	25	25	47	47	19	17
Environmental groups								
NI	25	37	53	46	11	6	3	2
GB	26	31	51	50	13	9	4	3
Government								
NI	3	3	46	48	32	32	11	10
GB	3	3	35	34	38	38	18	17
Ordinary people								
NI	10	10	53	53	25	25	4	2
GB	9	9	47	53	28	22	8	4

In addition, education and age were again found to be distinguishing factors in attitudes towards government, with the young and the highly educated being the least trusting of all. No clear trends were apparent with regard to trusting ordinary people, although the over 60s and Catholics tended to be the least trusting within their groups while those educated to degree level were the most trusting.

With respect to trust in scientists, attitudes in Northern Ireland have remained fairly consistent with almost 70 per cent showing some or a lot of trust, although there has been a slight increase in the direction of more trust. In GB, overall levels of trust in scientists have improved by some 6 per cent. In 1993, Yearley (1995) found only a slight social class effect with the highest classes being more trusting. In the 1996 survey, this effect is very much more evident. Over three-quarters of those in classes I and II and III non-manual had some or a lot of trust in scientists, compared to 65 per cent in III manual class and less than half in classes IV and V. The over 60s and the 18-30 year olds were less trusting than the other two groups. Also, Protestants and females were around 10 per cent more trusting than others in their categories. Most significant of all was the very clear trend with regard to level of education. Of those educated to degree level, almost 90 per cent have some or a lot of trust in scientists, falling to just over half of those with no formal qualifications. Overall trust in scientists remains high and they are consistently considered the most trustworthy when it comes to ranking public decision-makers (with politicians and journalists at bottom). Science is generally seen as the basis for 'progress' and although public opinion about the environment tends to fluctuate, there is a consistent reliance on science and independent analysis, possibly in the belief that it will find the true answers to serious environmental questions.

Of most interest in Table 6.6 are the levels of trust in business and industry. In GB, attitudes have remained virtually the same. In Northern Ireland however, even over the short three-year period, attitudes have changed quite dramatically with a very substantial decline in trust to levels very similar to those in GB (just over one quarter of those sampled). With the exceptions of the highest socio-economic class being most trusting and the 18-30 year olds being least trusting, the only factor displaying any sort of systematic pattern was educational qualification. The least trusting of all were those educated to degree level, with almost 80 per cent showing very little or no trust compared to just over half for those with no qualifications. In the absence of any recent major environmental events peculiar to Northern Ireland, no obvious explanation of the decline in trust comes to mind. More generally though, the prominent high profile of environmental issues in the media may have had an influence. Coverage of events like the 'Sea Empress' oil spillage, the campaigns against road and airport construction as well as the environment/quality of service versus profits debate in newly privatised industries, may have contributed to a fundamental shift in public attitudes towards business and industry. The fairly dramatic change in opinion in Northern Ireland for no locally determined reason suggests that the trickle-down effect predicted by Stringer (1992) might be taking place.

Respondents in both surveys were more trusting of environmental groups making the right decisions than any other group tested. Over a three-year period, Northern Ireland and GB both show a positive increase in the levels of trust. In the Northern Irish case, there has been a 12 per cent increase in the 'a lot' category compared to only 5 per cent in GB. This dramatic shift in opinion almost directly mirrors the negative change in attitudes toward business and industry, a trend not experienced in GB. In examination of the Northern Irish data, only class and education display any logical trends with differences of opinion varying most noticeably at the lower scale levels, social classes IV and V and those with no formal qualifications being the least trusting of all (13 per cent and 11 per cent respectively with little or no trust). Again no obvious reason explains the change, although it is not all that unreasonable to expect an increase in support for pro-environmental organisations with a hardening of attitudes towards those who are generally perceived as the greatest threat i.e. business and industry. Greater awareness of (but not necessarily knowledge of) *global* environmental issues such as global warming and ozone depletion, combined with media exposure of the various groups now involved in direct action campaigns, has done a lot to raise public awareness and concern. Major victories such as that of Greenpeace over Shell in the Brent Spar affair may well help reinforce the voluntary group/business divide.

Environmental Protection

A clear trend has emerged from the previous section demonstrating that the Northern Irish public place relatively more trust in the government on environmental issues than the British public. So far, attitudes towards the concepts of responsibility and trust have been examined considering the key players independently. It is useful to extend the analysis to incorporate some more difficult real world choices where conflicts of interest may arise between various interested parties. Respondents were asked whether they think government should create laws to protect the environment or whether ordinary people and businesses should be allowed to decide for themselves how to protect the environment, irrespective of the fact that they may not always do the right thing (Table 6.7).

The vast majority of people believe that government should pass laws to make businesses protect the environment while opinion is less enthusiastic in supporting the notion of similar laws being applied to ordinary people. Comparing the two surveys, in GB more respondents tend to favour laws directed towards business (7 per cent more) while in Northern Ireland just slightly more are in favour of laws directed at the ordinary citizen (4 per cent more).

These figures are interesting if we consider Table 6.1 where around 90 per cent of people believed it was the government's responsibility to make industry do less damage to the environment. It is not possible to make a direct comparison between the two statements given that Table 6.4(b) refers to businesses while Table 6.1 refers to industry. People may simply be making a distinction between the two. It is equally, if not more, plausible that the latter question introduces the stronger concept of enforced behaviour and restriction in the right to make our own decisions, which some people may object to.

Further examination of the Northern Irish data reveals that the patterns already identified above also prevail on these issues, with socio-economic status and educational qualifications having the most obvious effect on people's opinions. The higher the social class and the higher the level of educational qualifications the more favourable people are towards government passing laws to make business and ordinary people protect the environment.

Respondents were also asked to make a choice between the statements that a) It's mainly up to the government to protect the environment as ordinary people can't do much on their own or b) It's mainly up to ordinary people to do what they can to protect the environment, as the government can only do a limited amount. It should be borne in mind that this is a difficult question to interpret as the term 'a limited amount' is quite ambiguous and could be interpreted in many different ways by different people. In Northern Ireland, opinions were evenly split with 49 per cent choosing a) and 48 per cent choosing b). The corresponding figures for GB were 54 per cent and 43 per cent respectively, demonstrating a more common belief in government being more capable and responsible for protecting the environment than ordinary people. The Northern Irish tend to have a slightly more balanced view that government and ordinary people are equally responsible and capable of protecting the environment. In reference to Table 6.4, it appears that the Northern Irish respondents were slightly less inclined to believe that they themselves, as individuals, are as capable of protecting the environment as 'ordinary people' (collectively) than British respondents.

Table 6.7   Closest to respondent's view

	NI %	GB %
a)		
Government should let *ordinary people* decide for themselves how to protect the environment, even if it means they don't always do the right thing	22	24
OR		
Government should pass laws to make *ordinary people* protect the environment, even if it interferes with people's rights to make their own decision	52	48
b)		
Government should let *businesses* decide for themselves how to protect the environment, even if it means they don't always do the right thing	7	4
OR		
Government should pass laws to make *businesses* protect the environment, even if it interferes with business' rights to make their own decision	76	83

In GB, increased government spending is less well supported than in Northern Ireland (Table 6.1), despite the fact that a higher proportion view government as being responsible for environmental protection. The British though have a greater tendency to favour government action in the form of laws (Table 6.7) which force business and industry, in particular, to function in an environmentally friendly way, thus reducing the need for direct spending and the possibility of increased taxation.

This point is supported by the response to a more direct statement on the possibility of government making financial contributions. Respondents were asked whether they thought the government should help factories meet the costs of preventing pollution or whether those factories that cause pollution should pay the bills themselves. The Northern Irish were 5 per cent more inclined to accept the first option (25 per cent in all) than their British counterparts. There appears to be a fairly consistent trend that, while overall attitudes may be very similar between the two countries, the Northern Irish tend to be more in favour of direct and increased government financial intervention on environmental issues.

Costs of Protecting the Environment

It has been interesting so far to examine attitudes to general issues of environmental concern and responsibility. There is already some evidence suggesting that people's attitudes change when it comes to weighing the costs of protecting the environment against, for example, the impact on freedom to make decisions and increased government spending. It is worthwhile continuing this line of enquiry to investigate further how attitudes may change when various other cost factors are introduced as the trade-off for environmental protection. Table 6.8 presents the responses for both Northern Ireland and GB to four statements concerning definitive pro-environment actions of the key players (government, industry and the public) couched in terms of the potential costs (higher taxes, lower profits and fewer jobs, higher prices and restricted car use).

Both Northern Ireland and GB have a similar majority which agrees that the government should do more to protect the environment even if it means higher taxes. The much higher percentages favouring this statement compared to those in Table 6.1(a) suggest that there is greater support for more focused financial investment in environmental protection *per se* than just simply the environment. For Northern Ireland the figures represent a very slight increase on the previous year but for GB it means a 4 per cent decrease in support of the statement. This might well be a reflection of greater concern on the mainland with general Conservative taxation policies at the time. The figures do tend to confirm other findings already discussed, which reveal a much greater decline in support on the part of the British public towards increased government spending during the 1990s compared to Northern Ireland. It may also be partly due to the fact already mentioned that in GB local government has more responsibility for environmental matters and the possibility of paying through increased taxation at the local government level may be seen as a consequence of increased spending. For Northern Ireland, financial decisions on issues of taxation and the environment are taken at central and regional government levels respectively and as such do not constitute a direct threat of added cost in terms of local district council rates. For example, whereas local authorities in the rest of the UK already have a leading role in implementing environmental legislation such as the Local Agenda 21 programmes, most local authorities in Northern Ireland are just beginning to make plans and create programmes. It remains to be seen if this changing situation, albeit gradual, could have an effect on public attitudes, even in the near future.

A similar proportion of respondents (over 65 per cent) in both Northern Ireland and GB either agreed or strongly agreed with the statement that industry should do more to protect the environment, even at the cost of profits and jobs. What is most interesting though is the change in attitudes from the previous year. Northern Ireland experienced a 10 per cent increase in these combined categories and GB experienced a 6 per cent increase. This is a particularly difficult question to evaluate though as no distinction is made between lower profits (which would directly affect businesses) and fewer jobs (which would affect individuals and the local population). For Northern Ireland at least, the results complement earlier findings of a sharp fall in levels of trust for business and industry to make the right decisions about the environment (Table 6.6). It is quite clear that in Northern Ireland, the continued strong belief in government's responsibility for, and increased spending on, the environment has been accompanied by a quite remarkable hardening of attitudes towards business and industry.

Both surveys show a majority (approximately 60 per cent) in agreement or strong agreement that ordinary people should do more to protect the environment even if it means higher prices. Interestingly, these figures represent a fall in levels from the previous year (4 per cent in Northern Ireland and 7 per cent in GB). Overall these figures suggest a swing of opinion in favour of environmental protection, but not at the cost of the individual.

The more romantic notion of protecting the environment at all costs is rapidly giving way to the more practical reality that such actions have their price. Paying the price through increased government spending invariably means more taxes. This is a point viewed less enthusiastically by the British public than the Northern Irish for reasons mentioned above.

It is also apparent that the public is much less likely to support actions which impinge on the individual's rights to decide what to do or which have a direct financial cost. This point is further supported by the responses to the statement on car usage where a much lower proportion (44 per cent in Northern Ireland and 47 per cent in GB) favoured the pro-environmental option. Only around 15 per cent though, in both countries, actually disagreed or strongly disagreed with the contention that people should be allowed to use their cars as much as they like, even if it caused damage to the environment.

Table 6.8   The costs of protecting the environment

a) The government should do more to protect the environment, even if it leads to higher taxes.

	Strongly agree	agree	neither	disagree	strongly disagree
	%	%	%	%	%
NI	9	47	32	8	1
GB	12	45	29	11	1

b) Industry should do more to protect the environment, even if it leads to lower profits and fewer jobs.

	Strongly agree	agree	neither	disagree	strongly disagree
	%	%	%	%	%
NI	11	56	23	5	1
GB	17	50	24	7	0

c) Ordinary people should do more to protect the environment, even if it means paying higher prices.

	Strongly agree	agree	neither	disagree	strongly disagree
	%	%	%	%	%
NI	9	50	27	10	1
GB	9	51	26	11	1

d) People should be allowed to use their cars as much as they like, even if it causes damage to the environment

	strongly agree	agree	neither	disagree	strongly disagree
	%	%	%	%	%
NI	1	16	35	38	6
GB	2	14	35	37	10

In the breakdown of the Northern Ireland responses by socio-economic indicators, very similar trends to those already mentioned above were evident. Socio-economic status displayed a consistent trend across all of the statements with the highest classes expressing the strongest pro-environmental attitudes and the lowest classes the least. The class effect was most evident in responses to the statement about people with almost a

30 per cent difference in agreement between the highest and lowest categories on whether people should do more to protect the environment. A fairly consistent trend emerged with respect to education although the most noticeable differences were between the top and bottom categories. With only a few minor exceptions, those with degrees were very much more pro-environmentalist while those with no formal qualifications were the poorest supporters of all.

The effect of age displayed a consistent pattern throughout with the over 60s being the least supportive of more effort to protect the environment. Interestingly, they were followed by the 18-30 year olds who tended to be more indecisive and less agreeable to the environmentalist options than other groups. Place of residence showed no real discernible pattern while females and those categorised as other than Catholic or Protestant tended to be more pro-environment.

Conclusion

In this chapter, an attempt has been made to examine attitudes towards the environment in Northern Ireland focusing, in particular, on aspects of responsibility, trust and action relating to environmental protection. Temporal change within the province since 1990 and a spatial comparison with GB formed the main lines of investigation from which a number of noteworthy conclusions can be made. In general, opinions in Northern Ireland towards most basic environmental issues are quite favourable and recent trends suggest a gradual but continued improvement throughout the 1990s leading towards a levelling out of previously quite disparate attitudes between Northern Ireland and GB. From the environmentalist's perspective in the province, it must be encouraging to see that, even compared to GB, more people in Northern Ireland are trying (or at least claim to be trying) to do what is right for the environment in spite of a general tendency to be less confident in the capability of the individual to influence environmental matters. Public opinion is also more likely to support the notion of increased government spending than in GB.

There is some evidence to support the idea of a trickle-down effect, although the pattern and nature of change is not always as straightforward or apparent as the theory expressed by Stringer (1992) might suggest. Most noticeable in terms of overall change though, is the dramatic decline in levels of trust (by almost 10 per cent) in business and industry within Northern Ireland to levels more comparable with those in GB, which have hardly changed at all. The extent of change in attitude is quite extraordinary given that it occurred over a very short period of time (3 years) and for no immediate or apparent reason. Mirroring this and almost equally striking is the increase in levels of trust in environmental groups to match currently expressed opinion in GB. In contrast though, for some specific issues (e.g. levels of government spending on the environment) the closing gap in attitudes is a consequence of a fairly dramatic change of opinion in GB rather than in Northern Ireland.

There appears to be a growing perception that protecting the environment may have important cost implications. The environmental good will of the 1980s is now tempered with a rational understanding that protecting the environment may often involve a trade-off in terms of financial costs and possible restrictions on our personal rights. This is reflected in less enthusiastic support for environmental principles when such factors are proposed as competing options. It is also most apparent when the costs involved are either direct costs to the individual or indirect costs such as taxation. For instance, in spite of very popular overall support in both countries for the contention that government should be responsible for environmental protection, there is a tendency to prefer a legislative and law making option, targeted particularly at business and industry, to a more direct spending option. Respondents in GB tend to be more likely to adopt this attitude than those in Northern Ireland for reasons discussed above. Evidence even suggests that in both countries there is a slight reduction in the overall positive responses to the environment-friendly options in statements incorporating various cost factors.

Considering the effects of social and economic factors on attitudes, socio-economic status, level of educational qualifications and, to a lesser extent, age, were the most discriminatory across all of the issues examined. Certain patterns found in previous surveys were also evident. In particular, class and education factors tended to have a greater influence on attitudes in Northern Ireland than in GB. In general, those in the older age groups, those with low educational qualifications and those in lower socio-economic groups were the least likely to support pro-environmentalist options, although in many cases, majorities were supportive of them. Stringer (1992) suggested that the trickle-down effect would probably be more evident in the educational dimension than the other socio-economic categories, mainly on account of the greater ability of the better educated to understand the intellectual complexities of many underlying issues related to environmental matters. He used length of educational experience in his study, whereas this study adopted level of educational qualifications obtained; a direct comparison is not really possible although both indicators are probably very well correlated. In a general comparison with the 1990 survey though, education certainly appears now to play a much more prominent role as a discriminating factor. The most extreme differences in attitudes were frequently recorded between the highest and lowest educationally qualified groups, attitudes towards governmental and individual responsibility for the environment being typical examples.

On occasion, attitudes were observed to vary according to place of residence, religion and gender but trends were not as consistent or as significant as identified for the other variables. Females did tend to be slightly more supportive of general environmental issues than males and while Protestants were more trusting than Catholics, especially towards government, those classified as 'other' were more likely to be pro-environment in both opinion and action. These observations for religion generally concur with those identified in the 1990 (Stringer, 1992) and 1993 (Yearly, 1995) surveys. Relative to the 1990 survey though, religion tends to feature less prominently as a predictive variable for attitudes compared to the other factors considered.

References

Bryson, I., Devine, P. and Dowds, L. (1997), 'Transport in Northern Ireland: Finding the Way Forward' in L. Dowds, P. Devine, and R. Breen, (eds), *Social Attitudes in Northern Ireland: The Sixth Report*, Appletree Press, Belfast.

Christie, S. (ed), (1996), *Environmental Strategy for Northern Ireland* Northern Ireland Environment Link, Belfast.

Department of Economic Development. (1993), *Growing a green economy*, DED, London.

Department of the Environment. (1996), *Indicators of Sustainable Development for the UK*, HMSO, London.

HMSO. (1994), *Sustainable Development, the UK Strategy*, HMSO, London.

Lowe, P. and Goyder, J. (1983), *Environmental Groups in Politics*. George Allen and Unwin, London.

McGrew, A. (1993), 'The Political Dynamics of the 'New' Environmentalism', in D. Smith, (ed), *Business and the Environment; Implications of the New Environmentalism*, Paul Chapman Publishing, London.

O'Riordan, T. (ed) (1995), *Environmental Science for Environmental Management*, Longman, Harlow.

Smith, D. (ed) (1993), *Business and the Environment; Implications of the New Environmentalism*, Paul Chapman Publishing, London.

Stringer, P. (1992), 'Environmental Concern', in P. Stringer, and G. Robinson, (eds), *Social Attitudes in Northern Ireland: The Second Report*, Blackstaff, Belfast.

Witherspoon, S. (1995), 'The Greening of Britain: Romance and Rationality', in R. Jowell, J. Curtice, A. Park, L. Brook and K. Thomson, (eds), *British Social Attitudes the 11th Report*, Dartmouth, Aldershot.

Yearly, S. (1995), 'Environmental Attitudes in Northern Ireland', in R. Breen, P. Devine and G. Robinson, (eds), *Social Attitudes in Northern Ireland: The Fourth Report*, Appletree Press, Belfast.

7 Belief and Trust in the Political Process

MARTIN MELAUGH

Introduction

This chapter looks at the responses to a range of questions designed to obtain the views of respondents on the working of the political process and their trust in it. The module of questions on matters relating to political efficacy and political trust was also included in the 1994 Northern Ireland Social Attitudes (NISA) survey and a chapter on efficacy and trust, based on the responses to questions in that particular module, appeared in the fifth volume of *Social Attitudes in Northern Ireland*. That chapter, 'Public Support for Democratic Values in Northern Ireland' (Hayes and McAllister, 1996), contained a comprehensive introduction to the concepts behind political efficacy and trust, in addition to an analysis of the data. The information in this current chapter is intended to complement the earlier analysis of the 1994 data.

One of the central themes of the 1997 United Kingdom (UK) general election was the notion of trust. It could be argued that the Labour Party won that election not merely because the party had undertaken fundamental changes to its core policies, but also because the leadership of the party was able to persuade the British electorate that it could be trusted to form the government of the UK. Within Northern Ireland the question of trust in the political process is also an important one. Historically, the electorate of the region was unable to vote for any of the parties likely to form the UK Government and so returned representatives of local parties to Westminster. However, many issues relating to people's opinion of political parties, institutions, and the process of government, are similar in nature to those in Britain. Northern Ireland does of course have many unique social, economic and political characteristics which introduce different perspectives on the question of trust in politics and politicians. Some of these differing perspectives on political efficacy and trust in Northern Ireland are considered below.

Before turning to the analysis it is worth saying a word about Northern Ireland surveys in general and the NISA survey in particular. Those who are familiar with Northern Ireland surveys will be aware of some of the problems that social scientists in the region tend to encounter. There is, for example, the discrepancy between the responses people give in surveys, particularly on matters related to politics and religion, and how they subsequently behave. It is obvious that a section of respondents do not provide truthful answers to particular questions. Among these are questions designed to gauge whether the respondent holds moderate or more extreme political views. This results in a tendency, for example, for the political strength of the Alliance Party of Northern Ireland to be overestimated while that for Sinn Féin is usually underestimated (Whyte, 1990). Thus, for example, in the 1996 NISA survey only 2 per cent of respondents stated that they supported Sinn Féin, whereas the party obtained 15.5 per cent of the total vote at the 1996 Forum Election (Downey, 1996).

Religion remains a key classification variable in most social surveys in Northern Ireland. Even a cursory look at the information presented in the other chapters of this book will show this to be the case in the present survey. The question in the 1996 NISA survey dealing with religion produced a sizeable number (14 per cent) of respondents who answered that they were of 'no religion'. This is an outcome common to many surveys in Northern Ireland and has also become a feature of responses to the religion question in the census since 1971. It is unlikely, given the still conservative nature of the society and the relatively high level of religious observation, that Northern Ireland contains such a proportion of genuine atheists or agnostics. It is more likely that people use the 'no religion' category, or the 'none' category in the case of the census, as a means of politely refusing to answer the question. For the purposes of the analysis in this chapter the religion variable is based on the denomination in which the respondent was raised, that is, their family's religion. The justification for doing this is that many of the opinions held by people were formed when they were part of a particular community and many aspects of their circumstances, for example education and employment, were determined to a large extent by the community from which they came. McGarry and O'Leary (1995) use the terms 'cultural Catholics', and 'cultural Protestants', to refer to both those who are practising Catholics, or Protestants, as well as those who are born into the Catholic or Protestant religion but no longer believe or practise its tenets. The main impact of this categorisation on the present data was to reduce the number of cases in the 'no religion' category to 2 per cent.

In addition to questions on religion respondents were asked; 'generally speaking do you consider yourself as a unionist, a nationalist or neither?' The results of respondents political identification, broken down by family religion, is provided in Table A7.1 in Appendix A. Fifty-seven per cent of Catholics described themselves as 'neither' compared to 32 per cent of Protestants. Respondents were also asked; 'do you think the long-term policy for Northern Ireland should be for it to remain part of the UK, or to unify with the rest of Ireland?' Table A7.2 in Appendix A contains the responses to this question categorised by political identification. It is evident from the table that there was a certain degree of 'cross-over' between the categories, with 4 per cent of those who classified themselves as 'unionist' considering a United Ireland to be the best long-term policy, while 18 per cent of 'nationalists' felt that the best policy was for Northern Ireland to remain part of the UK. For those unaccustomed to Northern Ireland politics, the fact that someone can categorise themselves as a nationalist and yet support the union, or as a unionist and yet indicate a preference for a United Ireland, may seem to be contradictory. However, these results are a reminder that the way in which people identify themselves politically can differ from what they pragmatically consider to be the best way forward for the region.

Finally it is worth noting that in the course of the analysis it has been necessary to 'collapse' some of the categories of particular variables. This was done because the number of cases in particular categories were so small as to make assessments based on the data unreliable and certain statistical tests invalid. The amalgamation of categories was only undertaken where they could be merged in a meaningful way. This chapter is only intended to highlight some of the areas of interest in the data and is merely indicative of the type of analysis that could be performed on the responses to the module of questions on political efficacy and trust.

Community Differentials in Political Efficacy and Trust

The 1996 NISA survey contained a series of questions which were designed to gauge firstly, the extent to which respondents felt they had some influence over the political process, and secondly, the level of trust they had in elected representatives and governments. These two issues, termed political efficacy and political trust, are associated but distinct concepts (Craig, *et al*, 1990). Political efficacy relates to whether or not individuals believe that they have, or can have, an impact on the political process. More recently this concept was shown to contain two elements. The first being the ability of an individual to understand and participate in political life and the second being the belief that political representatives are responsive and responsible to the electorate (Hayes and Bean, 1993). Measures of political trust are usually designed to try and determine the extent to which people believe that public officials and political representatives act in the best interests of society, rather than in their own interests or the interest of a particular political party.

Both the concepts of efficacy and trust are believed to be central to the maintenance of a healthy democratic society. Members of such a society must believe that they can have some influence on the political process and must also have trust in those who are elected or appointed to positions of power within that process. There must also be active involvement in the political life of a democratic society with people being willing to support a range of activities that make the process operate. The absence of these conditions, that is where there are high levels of powerlessness and mistrust, and where people refuse to become involved, is believed to pose a danger to the stability and the very survival of any democratic state. Paradoxically, it is argued that too high a level of involvement in the political process can also have a destabilising effect.

Since the idea of political efficacy was first explored (Campbell, *et al*, 1954) the items used to measure the concept have undergone a number of refinements. As mentioned above respondents were asked a series of questions which gauge both efficacy and trust. Most of these questions were identically worded to those used in the 1994 NISA survey thus enabling comparisons to be made. Table 7.1 contains the results from a selection of the questions asked. The first five questions deal with issues related to efficacy while the next five questions relate to trust. In addition to Northern Ireland results for 1996 and 1994, Table 7.1 also provides a breakdown of the latest figures by religion and the results of identical questions from the 1996 British Social Attitudes (BSA) survey.

A comparison of the Northern Ireland results between 1994 and 1996 shows that in the case of nine out of ten of the items there has been a decrease in political efficacy and trust. The exception was an increase in the proportion of people disagreeing with the statement 'people like me have no say in what the government does'. One possible explanation for the increase in this particular item is that an election to the Northern Ireland Forum was held on 30 May 1996 which coincided with the period of fieldwork for the 1996 NISA survey. At that election 65 per cent of the electorate opted to have a say in who would represent them in the Northern Ireland Forum and in the multi-party talks at Stormont. Much of the decline in efficacy and trust between 1994 and 1996 is of the order of two or three percentage points on most of the items. While these differences are probably not statistically significant, taken together they represent an important change over a relatively short timeframe.

Table 7.1 also includes 1996 figures for the whole of GB (England, Scotland, and Wales) which allows a comparison with the 1996 data for Northern Ireland. In the case of seven of the questions the percentage scores were lower, sometimes by a large margin, in Northern Ireland than in GB. On two of the questions the figures in each area were the same. The aggregate figures for GB and those for Northern Ireland undoubtedly conceal important differences within each area. Hayes and McAllister (1996) gave a breakdown of British figures for 1994 which showed important differences between England, Wales, and Scotland. Each of the constituent countries of the UK has its own unique historical characteristics and differing political and social landscapes. It is not surprising therefore that there are some differences in the range of democratic values of the people that live in those areas.

Within Northern Ireland there are also important differences in political efficacy and trust between those respondents raised in the Catholic community and those in the Protestant community. It is apparent from Table 7.1 that in eight out of the ten items Catholics gave responses which indicated lower levels of efficacy and trust than Protestants. In the case of two of the five questions related to political efficacy, and four of the five items on political trust, these differences were statistically significant. It is worth noting at this point that the question 'How much do you trust police not to bend the rules in trying to get a conviction?' is the most problematic item in Table 7.1 for the current analysis. The question was included because it appeared in the British survey and it therefore can be used for comparative purposes. In a Northern Ireland context however the question of trust in the police is distinctly different from, say, that of trust in local councillors. The item does of course offer a genuine insight into one aspect of trust in the police service in Northern Ireland. Nevertheless, given the real difference in the level of support for the Royal Ulster Constabulary (RUC) in the two main communities in the region, this question was certain to produce a large divergence of opinion.

Thirty-two per cent of Catholics expressed trust in the police not to bend the rules compared with 55 per cent of Protestants, resulting in the largest difference (23 per cent), in terms of religion, in any of the items in Table

7.1. This point is worth bearing in mind when considering the section below which looks at the results of an index of efficacy and trust.

Table 7.1   Belief in the political process and trust in political representatives by religion 1996, with comparative data

	1996 Cath.#	1996 Prot.#	1996 NI	1994 NI	1996 GB~
	('disagree' or 'disagree strongly')				
	%	%	%	%	%
Political efficacy					
People like me have no say in what the government does	11*	21*	16	14	24
Sometimes politics and government is so complicated that a person like me cannot really understand what is going on	17	14	15	18	23
Generally speaking those we elect as MPs lose touch with people pretty quickly	9	9	9	11	11
It doesn't really matter which party is in power, in the end things go on much the same	14*	23*	18	22	31
Parties are only interested in people's votes, not their opinions	7	10	9	14	12

	('almost always' or ' most of the time')				
	%	%	%	%	%
Political trust					
UK governments of any party to place the needs of the nation above the interest of their own political party?	12*	26*	19	22	23
Politicians of any party in the UK to tell the truth when they are in a tight corner?	6*	11*	9	10	9
Local councillors of any party to place the needs of their area above the interests of their political party?	27	33	30	34	29
Top civil servants to stand firm against a minister who wants to provide false information to parliament?	25*	36*	31	33	31
Police not to bend the rules in trying to get a conviction?	32*	55*	45	48	53

for the purposes of the analysis in this chapter the religion variable is based on the denomination in which the respondent was raised, that is, their family's religion
~ GB is defined as England, Scotland, and Wales
* denotes that the difference between religious denominations is statistically significant at p<0.05

Those who know something of the history of Northern Ireland would not find the differing responses between Catholics and Protestants in Table 7.1 surprising.   The experience of the political process by the two main communities in Northern Ireland has been radically different since the establishment of the state. The Protestant community had a vested interest in the political survival of Northern Ireland and in many respects Protestants were rewarded for their support. Catholics were antagonistic towards partition and many took no part in the various institutions that made up the structure of the state, while others were excluded from these institutions by discrimination. The in-built unionist majority in the six counties that formed Northern Ireland, assured 50 years of one-party rule with no possibility of a change of government. Nationalists felt alienated from the Stormont parliament and had little power and less trust in the operation of government. In more recent times a number of developments, including the replacement of Stormont by direct rule from Westminster and the imposition of the Anglo-Irish Agreement, have all had an alienating effect on Protestants (Dunn and Morgan, 1994).

Other Socio-economic Characteristics of Respondents

Table 7.1 contains results broken down by the religious background of respondents. This section briefly considers the effects of a range of other variables. These variables are: gender; age category; residential area; education; social class; political identification; and party partisanship. To allow a parsimonious presentation the results of this analysis are discussed only in outline. For example, it would be possible to summarise the results by saying that for two of the five efficacy items and for four of the five trust items, there were statistically significant differences between Catholics and Protestants. A similar summary approach is used now for the other explanatory variables.

Significant differences between male and female respondents are apparent in the case of two of the political efficacy items and two of the trust items. Women (91 per cent) were much more likely than men (77 per cent) to state that they found the political process 'so complicated that a person like me cannot really understand what is going on'. Women (17 per cent) were also significantly less likely to express trust in the UK government than men (23 per cent). There was only one efficacy item and two trust items where the age category of a respondent proved to be significant. Older people were more likely to agree that politics and government were sometimes too complicated to understand. Those in the oldest age band, 65 years and over, were more trusting of the police and, along with the youngest age band, 30 years and under, were more likely to trust civil servants. On three of the efficacy items and on three of the trust items there were statistically significant differences depending on area of residence; Belfast, rural areas, or urban areas outside of Belfast. In the case of the trust items the pattern was clear, in that those from a rural area tended to be much more trusting than those from either Belfast or from other urban areas. There was no similar pattern with the items on efficacy.

For ease of analysis the original categories of educational attainment were merged into four groups: degree level; 'A' level; 'O' level; and those with no qualifications. Respondents with higher levels of education were significantly more likely to feel that they understood the political process compared to those with lower levels of education, and were also significantly less likely to feel that it didn't matter which party was in power. However, on other questions there was a less clear-cut distinction between higher levels of education and higher levels of political efficacy. In contrast with efficacy there was less association between education and political trust with only two of the five items proving to be statistically significant.

The initial five categories of social class were merged into three; again this was done to facilitate analysis. There were no significant differences between these new categories on any of the questions on trust but there were differences on two of the efficacy items. The lower a person's social class the more likely they were to agree that they had problems understanding politics and government. The same relationship was true for the statement that 'it doesn't really matter which party is in power'.

In the case of political identification there was a significant difference between unionists (22 per cent) and nationalists (13 per cent) who disagreed with the statement 'people like me have no say in what the government does'. None of the other four efficacy items showed statistically significant differences. However, in the case of the questions related to political trust, four of the five items showed significant differences with unionists reporting higher levels of trust than nationalists or those who described themselves as 'neither'. Given the strong association between religion and party partisanship (with the exception of the Alliance Party of Northern Ireland, APNI) it is not surprising to find that the pattern of significant results followed those found for family religion. There was one item under efficacy and four under trust where there was a distinct difference in the responses when broken down by the political party the respondent felt closest to. The Alliance Party (APNI), the Ulster Unionist Party (UUP), and the Democratic Unionist Party (DUP), were all more likely to disagree with the statement that 'it doesn't really matter which party is in power', than either the Social

Democratic and Labour Party (SDLP) or Sinn Féin (SF). (Due to the low number of reported SF supporters among respondents the results for this party should be treated with caution). On the items of trust the general pattern was that the UUP and the DUP tended to be more trusting than the SDLP or SF. The responses of APNI supporters to questions related to trust did not follow a consistent pattern in relation to the other parties.

In summary, in terms of political trust the most important differences were to be found in the case of political identification and party partisanship; in addition to family religion as demonstrated in the detailed breakdown in Table 7.1. In the case of efficacy there were significant differences depending on level of education. The 1996 NISA survey also included some questions on how much trust respondents felt in various potential governments under three scenarios; direct rule from Britain, devolved government at Stormont, and under a United Ireland. An analysis of the responses to these questions is provided by Niall Ó Dochartaigh in Chapter 4 of this book.

Index of Political Efficacy and Trust

Using the first five questions in Table 7.1 it was possible to construct a crude index of political efficacy. This was done by firstly assigning a score of '1' to those who disagreed with the statements, in other words those who believed that it was possible to have some influence over the political process, and a score of '0' to those who agreed with the statements. Then the scores from the first five items were added together. A similar procedure was used with items six to ten from Table 7.1 to produce a range of scores for political trust. The results for political efficacy and trust are contained in Table 7.2.

It is clear from Table 7.2 that almost 60 per cent of the Northern Ireland sample did not score anything on the five items that make up the measure of political efficacy. This is a particularly low score and the peripheral position of Northern Ireland in relation to the UK government at Westminster is likely to be part of the explanation. It is also evident that there were statistically significant differences between the aggregate scores for Catholics and Protestants. Catholics (63 per cent) were much more likely to have a score of '0' than Protestants (55 per cent). The level of political trust, as represented by the scores in the second half of Table 7.2, was higher than was the case for political efficacy. In contrast to efficacy, 'only' 35 per cent of people scored '0' on the cumulative index for political trust. To some degree this is explained by the particularly high level of trust in the police expressed by Protestant respondents. The breakdown by religion again showed a statistically significant difference in the responses of Catholics and Protestants. Almost half (47 per cent) of Catholic respondents scored '0' on the index compared to a quarter (25 per cent) of Protestants.

Table 7.2   Cumulative score of items on political efficacy and trust, by religion

	Cath.*	Prot.*	NI
	%	%	%
Score			
Political efficacy			
0	63	55	58
1	25	24	24
2	6	14	10
3	5	4	4
4	1	2	2
5	1	0	1
Score			
Political trust			
0	47	25	35
1	26	29	27
2	14	23	19
3	9	12	10
4	3	7	5
5	2	5	3

* denotes that the difference between religious denominations is statistically significant at $p<0.05$

It is possible to further summarise the results in Table 7.2 by dividing the scores in each section of the table into 'low' and 'high'. In order to allow for comparison with figures from 1994, the cut-off point adopted by Hayes and McAllister (1996) was used in the current analysis. This meant that those who responded favourably to at least two of the five items (scored 2 or more) were considered to have shown 'high' levels of political efficacy (17 per cent of respondents) or 'high' levels of trust (38 per cent of respondents). Those who did not meet this threshold (scored 1 or less) were categorised as being 'low'. The results from this breakdown for a number of sub-groups is contained in Table 7.3.

Table 7.3   Summary of political efficacy and trust, by sub-groups

	Political efficacy		Political trust	
	'low'	'high'	'low'	'high'
	%	%	%	%
Religion				
Catholic	88*	12*	73*	27*
Protestant	79*	21*	54*	46*
Gender				
Male	78*	22*	60	40
Female	87*	13*	64	36
Age				
Less than 30	82	18	58	42
31 to 50	81	19	62	38
51 to 64	86	14	69	31
65 and over	87	13	60	40
Residential area				
Belfast	83	17	67	33
Urban, outside Belfast	80	20	64	36
Rural	85	15	59	41
Education				
Degree/Higher education	76*	24*	48*	52*
'A' level or equivalent	73*	27*	62*	38*
'O' level or CSE	86*	14*	66*	34*
No qualifications	89*	11*	68*	32*
Social class				
Upper	77*	23*	59	41
Middle	84*	16*	60	40
Lower	89*	11*	69	31

* denotes that the difference between the categories is statistically significant at p<0.05

The figures in Table 7.3 indicate that there were some significant differences in the characteristics of those who scored low or high on political efficacy and trust. The results for religion provide a further summary of the information in Table 7.1, by showing a significant difference in terms of efficacy and trust, with Catholic respondents being much less likely than Protestants to be in the high category of both indices. Only 13 per cent of women were classified as showing high levels of efficacy compared to 22 per cent of men and these differences were statistically significant. The slight difference in the case of political trust was not significant. Although there were some differences across age categories, with younger people more likely to be classified as having a high level of efficacy, none of the differences, in either efficacy or trust were statistically significant. The area in which a respondent lived, whether in Belfast, urban areas outside of Belfast, or rural areas, made no significant difference to the summary measures of political efficacy and trust.

It is evident from Table 7.3 that the level of educational attainment was a significant factor in whether or not a respondent would score high on political efficacy and trust. The relationship in terms of trust was a linear one with higher levels of educational attainment being associated with higher levels of trust. In the case of efficacy there were distinct differences between those with no qualifications or low qualifications and those in the two higher categories. There were also marked differences in efficacy across the three categories of social class, with those in the higher classes being significantly more likely to display high levels of political efficacy. While those in the lowest social class showed less trust than those in the two higher categories, the difference was not significant.

The initial discussion of the relationship between the religion of a respondent and their level of political efficacy and trust demonstrated the importance of religious background. From the results in Table 7.3 it is apparent that some of the other variables were also related to efficacy and trust. Before moving to the next section it is worth briefly considering these relationships further. In particular, by looking at whether or not the independent variables had any effect on the relationship between religion and the summary scores of efficacy and trust. Table 7.4 includes the results for those who scored 'high' on efficacy or 'high' on trust and outlines what happens to any relationship between religion and efficacy or trust when the categories of another independent variable are considered separately.

It was demonstrated earlier, in Table 7.3, that there were significant associations between gender and efficacy. What is obvious from Table 7.4 is that within the categories of male and female there remained significant differences between the percentage of Catholic and Protestant respondents who scored high on efficacy. There was no initial significant difference between males and females on the level of political trust. However statistically significant differences did occur in the level of trust between Catholics and Protestants amongst both men and women.

Table 7.4   High political efficacy and trust, by religion, controlling for selected sub-groups

	Political efficacy 'high'			Political trust 'high'		
	Cath.	Prot.	NI	Cath.	Prot.	NI
	%	%	%	%	%	%
Gender						
Male	17~	27~	22*	27~	50~	40
Female	9~	16~	13*	28~	43~	36
Age						
Less than 30	11~	26~	18	30~	53~	42
31 to 50	16	22	19	26~	48~	38
51 to 64	6~	19~	14	22	36	31
65 and over	11	15	13	32	44	40
Residential area						
Belfast	12	24	17	14~	49~	33
Urban, outside Belfast	18	21	20	19~	43~	36
Rural	10~	21~	15	34~	49~	41
Education						
Degree/Higher education	21	27	24*	36~	63~	52*
'A' level or equivalent	21	31	27*	22~	49~	38*
'O' level or CSE	10	18	14*	29	38	34*
No qualifications	7~	14~	11*	24~	41~	32*
Social class						
Upper	20	25	23*	27~	50~	41
Middle	13	19	16*	29~	48~	40
Lower	6~	18~	11*	24~	40~	31

* denotes that there was a statistically significant (at p<0.05) difference between the high scores of efficacy and trust across the categories of the independent variables (that is, a summary of the results of Table 7.3)

~ denotes that there was a statistically significant (at p<0.05) difference between the high scores on efficacy and trust for Catholics and Protestants, after controlling for the independent variable

Whereas there were no statistically significant differences across the age categories of respondents, it is apparent from the breakdown in Table 7.4 that within particular age categories there were clear, and in some instances significant, differences between Catholics and Protestants. Much the same can be said in the case of residential area. While a respondent's place of residence did not provide any significant difference in terms of how they scored on efficacy and trust, within each residential area religion produced significant differences in terms of trust. Questions related to political efficacy also produced important differences but these were only significant for respondents from a rural area.

The earlier discussion of the impact of education showed strong differences in responses to efficacy and trust with each of the levels of academic achievement. The figures in Table 7.4 reveal that even though education was undoubtedly an important variable, the effect of religion was still strong across all attainment levels. Indeed in the breakdown for political trust there were significant differences between Catholic and Protestant respondents in three out of the four educational attainment categories. Similarly, in the case of social class the figures for trust were significantly different between Catholics and Protestants across the three summary categories. Finally, it is possible to summarise the discussion on Table 7.4 by saying that the results follow an overall pattern. In every category of every variable, there was a larger percentage of Protestants than Catholics who scored 'high' on both efficacy and trust. The results therefore show important differences between Catholics and Protestants regardless of differences in key personal and socio-economic characteristics.

Classification of Democratic Types

Having produced the summary variables of political efficacy and trust it was then possible to classify people according to a typology of democratic types described by Hayes and McAllister (1996, p.20). In this typology those who scored low on efficacy and low on trust were considered to be 'weak democrats'; not only did they not believe that they could affect the political process but they also had no trust in those who were part of that process. Any respondents who scored high on both efficacy and trust were considered to be 'strong democrats'; they believed in the political process and trusted those elected or appointed to political office. Those respondents who had a low score on efficacy but a high score on trust were termed 'optimists'; while they did not believe that they could influence the political process they were prepared to trust those who were part of that process. Finally, there were those respondents who scored high on efficacy but low on trust and were termed 'sceptics'; these people believed in the political process but had little trust in those in positions of power within that process. Using this classification Table 7.5 provides the results for Catholics and Protestants for 1996, the figures for Northern Ireland for 1994 and 1996, and the same information from the 1996 BSA survey.

Table 7.5   Democratic types in Northern Ireland, by religion 1996, with comparative data

	1996 Cath. %	1996 Prot. %	1996 NI %	1994 NI %	1996 GB %
Democratic type					
Weak democrats	65	45	54	49	48
Sceptics	7	9	8	9	12
Optimists	23	33	29	31	23
Strong democrats	5	13	9	11	16

The overall pattern of results between the aggregate figures for Northern Ireland and GB were roughly similar even though the details do differ. Those defined as 'weak democrats', that is those who scored low on both political efficacy and political trust, accounted for approximately half of respondents in both regions. The next largest group were those termed 'optimists', that is those with a low level of efficacy but a high level of trust, who numbered roughly a quarter of respondents. However, within this overall pattern Table 7.5 shows some important differences between the aggregate figures in 1996 for Northern Ireland and those for GB. There was a smaller percentage of people in Northern Ireland who were classified as 'strong democrats' (9 per cent) compared to Britain (16 per cent), and a higher level of 'weak democrats' (54 per cent compared to 48 per cent respectively). It is also obvious that there has been a marked change in Northern Ireland between 1994 and 1996. The percentage of those respondents classified as 'weak democrats' has increased from 49 per cent to 54 per cent. In summarising as it does all the information in Table 7.1, the results in Table 7.5 confirm the impression that there has been an important decline in the level of belief in democratic values in Northern Ireland in a relatively short period of time. The most significant differences in Table 7.5 were those between Catholics and Protestants. Of Catholic respondents 65 per cent were classed as 'weak democrats' compared to a figure of 45 per cent for Protestants. There were also many fewer strong democrats among Catholic respondents (5 per cent) than among Protestants (13 per cent). Among the group termed 'optimists' there is also an important difference between Catholics (23 per cent) and Protestants (33 per cent).

Table 7.6 provides a breakdown, by interest in politics, political identification, and party partisanship, of the summary scores on political efficacy and trust. Following the results on religion and given the close association between political identification and religion (see Table A7.1 in Appendix A), it is not surprising that there were important differences between nationalists (11 per cent) and unionists (19 per cent) in terms of efficacy. In the case of political trust the differences were most pronounced, and statistically significant, with 50 per cent of unionist respondents having a high score compared to the much lower figure of 19 per cent for nationalists. These figures very clearly reflect the differing Catholic and Protestant experiences, mentioned earlier, of the era of Stormont Government and the major political developments during the conflict over the past 30 years.

There were also statistically significant differences between respondents on the basis of which political party they would be most likely to support. Again these differences in the level of efficacy and trust reflect the high correlation between religion and support for the various political parties, with the exception of the Alliance Party of Northern Ireland (APNI) which attracts cross-community support. Among the other parties there were differences in the level of political efficacy between the Social Democratic and Labour Party (SDLP) (12 per cent) and the two Unionist parties, the Democratic Unionist Party (DUP) (20 per cent) and the Ulster Unionist Party (UUP) (21 per cent). The differences in terms of political trust were even wider with supporters of the SDLP (32 per cent) showing much lower levels of trust than those of the either the UUP (53 per cent) or the DUP (52 per cent). (The earlier note about the low number of reported Sinn Féin supporters among respondents still applies and the results for this party should be treated with caution). These figures, based as they are on political party partisanship, probably reflect the respondents perception of the historical role played by the parties in the political process in Northern Ireland. The unionist parties exercised power prior to 1972 and those who support them may express higher levels of political efficacy and trust partly as a consequence of experience of government. The SDLP is a comparatively modern party which, with the exception of the short-lived power-sharing Executive, has no experience of regional government and the low levels of political efficacy and trust on the part of those respondents who would support it probably reflects this situation.

Table 7.6 also shows the levels of efficacy and trust among respondents who declared an interest in politics ('some', 'quite a lot', or 'a great deal') and those who had little interest in the subject ('not very much interest', or 'none at all'). Significant differences were apparent in terms of both political efficacy and trust, with those showing some interest in politics being more likely to score highly on these measures. The results are consistent with findings in other similar surveys. Intuitively it is reasonable to assume that those who take some interest in politics are likely to consider that they understand the process, believe they can have some influence over it, and be more trusting of those who are part of the process.

Table 7.6   Summary of political efficacy and trust, by sub-groups

	Political efficacy		Political trust	
	'low'	'high'	'low'	'high'
	%	%	%	%
Interest in politics				
Some interest	79*	21*	59*	41*
Not very much interest	90*	10*	68*	32*
Political identification				
Nationalist	89	11	81*	19*
Unionist	81	19	50*	50*
Neither	83	17	64*	36*
Party partisanship				
Democratic Unionist Party	80*	20*	48*	52*
Ulster Unionist Party	79*	21*	47*	53*
Alliance Party NI	71*	29*	63*	37*
Social Democratic Labour Party	88*	12*	68*	32*
Sinn Féin #	79*	21*	85*	15*

* denotes that the difference between the categories of the relevant variable is statistically significant at p<0.05
# less than 20 cases

The results for level of interest in politics can also be further broken down by religion, as is the case in Table A7.3 in Appendix A. This table demonstrates that even among those who have some interest in politics there remained significant differences between Catholics and Protestants. Only 14 per cent of those Catholics who expressed some interest in politics had a high score on political efficacy compared with 26 per cent of Protestants. In the case of political trust the difference in the figures was even wider with 27 per cent of Catholics scoring highly as opposed to 50 per cent of Protestants. These results reflect a similar pattern to that found by Hayes and McAllister (1996). The results are important because those who express an interest in politics are considered a key constituency in any democratic society. It is from the ranks of this section of the population that political organisations depend for those participants who will actively support the political process. So, even among this crucial group, there is a higher percentage of Catholics than Protestants who feel that they have no influence over the political process and have no trust in the politicians and officials who are part of that process.

Conclusion

The 1996 NISA survey contained a module of questions which allowed an assessment to be made of the extent of political efficacy and political trust in Northern Ireland. Similar questions had been administered during the 1996 BSA survey and the 1994 surveys in GB and Northern Ireland. This permitted comparisons to be made of the results for Northern Ireland between 1994 and 1996, and also to briefly look at how Northern Ireland compared with GB.

There are many aspects of the political process, in terms of political experiences and levels of support for democratic institutions, which are common to each of the constituent countries of the UK. However, there are also important characteristics in each of these countries which make them distinct from each other. This is especially the case with Northern Ireland where the presence of a parliament at Stormont for fifty years and the political developments since direct rule in 1972, all help to give the region a unique political culture unlike anything else in the rest of the UK.

These particular political circumstances in Northern Ireland are reflected in the lower levels of political efficacy and political trust that were recorded for respondents in the region. The differences at a regional level are however not as great as the differences that were apparent between the two main communities in Northern Ireland. The views of Catholics and Protestants of the working of the political process and their trust in it are widely divergent. The historical divisions between those from the two main traditions, together with the legacy of contrasting experiences of unionist control of government at Stormont, and the different experiences of 30 years of 'the Troubles', have produced alternative views of the political process among Catholics and Protestants in Northern Ireland.

The results presented earlier also point to a marked decline in democratic values in Northern Ireland between 1994 and 1996. It is possible that this may simply be an aberration and future surveys in the region may show higher levels of efficacy and trust. There were certainly a number of factors around the time of the 1996 survey which may have influenced respondents. On the wider UK political scene there was a sense of disillusionment with politics as the media reported seemingly endless allegations of sleaze and corruption in political life. Within Northern Ireland the peace process appeared to have brought little real progress towards a political settlement and the Irish Republican Army (IRA) ended its 1994 ceasefire on 9 February 1996. These and other factors may have been, at least partly, responsible for the apparent decline in political efficacy and trust.

Democratic societies need a minimum level of active participants to make the system work properly and to sustain it over long periods of time. The figures for Northern Ireland are worrying because they indicate that a substantial section of the population, that is those with a background in the Catholic community, display very low levels of political efficacy and associated low levels of political trust. The high proportion of Catholic respondents who were classified as being weak democrats, and the corresponding low level of strong democrats, probably reflects the extent of political alienation that has historically been felt by that community.

There are plenty of indicators that large sections of the Catholic community remain apart from the political process in the region. The fact that the Social Democratic and Labour Party (SDLP), for example, still refuses to take an active role in the Police Authority of Northern Ireland, is one example of how much still needs to be done to achieve trust in a range of institutions within the Northern Ireland state. Nevertheless, there have been some positive aspects in terms of the Catholic community's involvement in the political process in Northern Ireland over the past 30 years. During a period of Northern Ireland's history when many people were turning their backs on politics in the region, the SDLP did manage to engage the Catholic middle-class in local government. And, regardless of one's opinion of Sinn Féin's (SF) political aims, there is no doubt that the party has developed a political infrastructure of supporters amongst a nationalist section of the Catholic population who traditionally took little part in the democratic process.

References

Campbell, A., Gurin, G. and Miller, W.E. (1954), *The Voter Decides*, John Wiley, New York.

Craig, S.C., Niemi, R.G. and Silver, G.E. (1990). 'Political Efficacy and Trust: A Report on the NEX Pilot Study Items', *Political Behaviour*, 12, pp. 289-314.

Downey, J. (1996), *Irish Independent*, Saturday 1 June, 11.

Dunn, S. and Morgan, V. (1994), *Protestant Alienation in Northern Ireland: A preliminary survey*, Centre for the Study of Conflict, University of Ulster, Coleraine.

Hayes, B.C. and Bean, C.S. (1993), 'Political Efficacy: A Comparative Study of the United States, West Germany, GB and Australia', *European Journal of Political Research*, 23, pp.261-280.

Hayes, B.C. and McAllister, I. (1996), 'Public Support for Democratic Values in Northern Ireland', in R. Breen, P. Devine and L. Dowds (eds), *Social Attitudes in Northern Ireland: The Fifth Report*, Appletree Press, Belfast.

McGarry, J. and O'Leary, B. (1995), *Explaining Northern Ireland*, Blackwell Publishers Ltd, Oxford.

Whyte, J. (1990), *Interpreting Northern Ireland*, Clarendon Press, Oxford.

Appendix A

Table A7.1   Political identification, by family religion

	Cath. (%)	Prot. (%)
Political identification		
Unionist	1	67
Nationalist	43	1
Neither	57	32

Table A7.2   Political identification, by respondent's view of best policy for Northern Ireland

	Remain part of UK (%)	Unify with rest of Ireland (%)	Other solution (%)
Political identification			
Unionist	95	4	1
Nationalist	18	69	13
Neither	62	25	13

Table A7.3   High political efficacy and trust, by religion and level of interest in politics

	'A great deal', 'Quite a lot', or 'Some' %	'Not very much interest', or 'None at all' %
Political efficacy: 'high' score		
Catholic	14~	10
Protestant	26~	11
Northern Ireland	21*	10*
Political trust: 'high' score		
Catholic	27~	28
Protestant	50~	39
Northern Ireland	41*	32*

* denotes that the difference between levels of interest in politics is statistically significant at $p<0.05$
~ denotes that the difference between religious denominations is statistically significant at $p<0.05$

Appendix I:
Technical Details of the Survey

ALAN McCLELLAND

Background to the Survey

The Social Attitudes survey in Northern Ireland (NISA) was funded by the Nuffield Foundation and the Central Community Relations Unit for the third consecutive year in 1991. Subsequent funding for the NISA survey was secured for a further four years (1993 - 1996) with contributions from government departments in Northern Ireland.

As in previous years, both the British Social Attitudes survey (BSA) and the NISA survey consisted of 'core' questions and of 'modules' on specific topic areas. Modules in the Northern Ireland questionnaire were selected from the larger number that were used in the BSA questionnaires. The one exception to this was a module dealing with issues specific to Northern Ireland which was included only in the NISA. However, some of these Northern Ireland module questions were, for comparative purposes, also asked in Britain.

An advisory panel consisting of representatives from Social and Community Planning Research (SCPR), the Northern Ireland Statistics and Research Agency (NISRA) and the Central Community Relations Unit (CCRU) were responsible for constructing the basic content of the questionnaire used in Northern Ireland. The panel both planned the Northern Ireland module and advised on which modules from the British questionnaire might be most usefully incorporated into the Northern Ireland version.

Final responsibility for the construction and wording of the questionnaire remained with SCPR. Responsibility for sampling and fieldwork rested with the Central Survey Unit of NISRA.

Since 1993, NISA survey fieldwork has been completed by interviewers employing Computer Assisted Interviewing (CAI) (see Sweeney and McClelland, 1994).

Content of the Questionnaire

The basic schema of the Northern Ireland Social Attitudes questionnaire mirrored that of the British survey. There were two components. The first consisted of the main questionnaire administered by interviewers. The second component was a self-completion supplement which was filled in by respondents after the interview and was either collected by interviewers or returned by post.

Table A1   Contents of the 1996 Northern Ireland Social Attitudes survey questionnaire

Core Questions	Newspaper Readership
	Government Spending
	Labour Market Participation
	Religion
	Classification
Topic Modules	Countryside / Environment
	Health Care
	Housing (long version)
	Political Trust
	ISSP Module (Role of Government)
	- self-completion only
Northern Ireland Module	Community Relations
	Perceptions of Religious Prejudice
	Protestant-Catholic Relations
	Segregation and Integration
	Equal Opportunities in Employment
	Education / Integrated Schools
	Political Partisanship
	Community / National Identity
	Party Political Identification

Each year the questionnaire includes a number of core questions such as the economy and labour market participation, as well as a range of background and classificatory questions. It also contains questions (or modules) on attitudes to other issues. These are repeated less frequently, on a two- or three-year cycle, or at longer intervals.

The self-completion supplement consisted of the module designed for the International Social Survey Programme (ISSP) (for further details of the ISSP see Jowell, *et al*, 1993), as well as items from the British questionnaire and the Northern Ireland module which were most appropriately asked in that format.

The Sample

As with the British Social Attitudes survey, the Northern Ireland survey was designed to yield a representative sample of all adults aged 18 and over, living in private households (for further details of the BSA see Jowell, *et al*, 1997).

The sample in Northern Ireland was drawn from the Valuation and Lands Agency (VLA) list in contrast to that in Britain which is based on the Postcode Address File (PAF) and involves a multi-stage sample design. The list provided by the VLA is the most up-to-date listing of private households in Northern Ireland and is made available to CSU for research purposes.

The VLA list available to CSU was limited to private addresses. It excluded people in institutions (but not those who live in private households at such institutions). Contained within the list, inevitably, were a proportion of 'non-viable' addresses which may have been, for example, derelict or vacant. The size of the allocated sample was adjusted to compensate for this wastage.

As the sampling frame was one of addresses, a further stage of sampling was required to select individual adults for interview. Consequently, weighting of the achieved sample was necessary to compensate for the effect of household size on the probability of individuals being selected as a respondent.

Sample Design and Selection of Addresses

Several factors common to Northern Ireland including the generally low population density outside greater Belfast and its small geographical area, allow the use of an unclustered, simple random sample design. In addition, the extensive coverage of CSU's field force enables this sample design to be effectively used. The benefits gained from using a simple random sample include: its effectiveness in generating representative samples of the population for surveys at any given sample size; and the greater precision of survey estimates compared to those of a clustered design.

The NISA sample was therefore a simple random sample of all private addresses contained on the VLA list. Addresses were selected from the computer-based copy of the VLA list using a random start, fixed interval selection procedure. Addresses selected for household surveys by CSU are excluded from further sampling for a period of two years. Prior to drawing the sample, Northern Ireland as a whole was stratified into three geographical areas. This stratification, based on district council boundaries, consisted of Belfast - Belfast district council; East - most of the remaining district council areas east of the river Bann, excluding Moyle and Newry and Mourne; West - the remaining district council areas. Within each of these three areas, a simple random sample of addresses was selected from the VLA list, with probability proportionate to the number of addresses in that stratified area. Figure A1 shows the distribution of addresses on the list, of selected addresses and the distribution of addresses at which interviews were achieved.

Selection of Individuals

The VLA list is a good up-to-date source of private addresses in Northern Ireland. The list does not however, include information about the number of adults living at each address. Only one individual was to be selected at each address at which interviewers were successful in achieving initial co-operation. To achieve this, the interviewers entered anonymised details of all the adults in the household currently aged 18 or over into the laptop computer. From the list of eligible adults, the computer selected one respondent through a Kish grid random selection procedure.

Figure A1   Geographical distribution of the sample

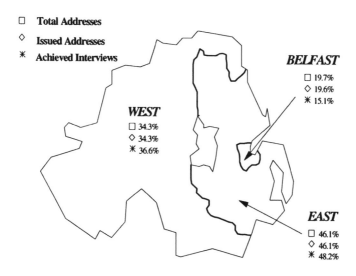

Weighting of the Achieved Sample

It is not possible using the rating list to select addresses in Northern Ireland with probability proportionate to the size of the household. To compensate for this potential source of bias, the data was weighted prior to analysis. The weighting procedure was further adjusted to compensate for a bias in the computer program which led to unequal selection probabilities of individuals within households.

The final weights were achieved by a two-stage process. The first stage was to derive unscaled weights for each of the households of different sizes. This was derived by dividing each of the total number of cases in households of each size by the number of cases of each of the person number groups who were selected for interview. For example, there were 399 households containing 2 persons who participated in the survey. This yields an unscaled weight of 2.714 for those who were selected respondent number 1 and an unscaled weight of 1.58 for those who were selected respondent number 2. In total, this process resulted in 15 unscaled weights.

Table A2   Weighting of the sample - stage 1

	Selected Respondent Number					
Number of persons in household	1	2	3	4	5	Total persons in households
1	239					239
2	147	252				399
3	40	46	9			95
4	2	15	12	6		35
5	2	7	-	2	3	14

Table A3   Weighting of the sample - stage 2

Weight No.		%	Scaled Weight
1	239	30.4	0.51
2	252	32.1	0.81
3	7	0.9	1.03
4	46	5.9	1.06
5	15	1.9	1.20
6	40	5.1	1.22
7	147	18.7	1.39
8	12	1.5	1.50
9	3	0.4	2.39
10	6	0.8	2.99
11	1	0.1	3.08
12	6	0.8	3.59
13	1	0.1	4.10
14	11	1.4	5.42

In order to retain the actual number of interviews, the very highest unscaled weights (11 cases) were trimmed with the remaining weights scaled back to the originally achieved sample size. This process produced 14 weights, yielding 786 interviews with an average weight of 1.

Fieldwork

Prior to commencement of the fieldwork, advance letters were sent to each household selected in the sample. The letter informed the household that they had been selected for inclusion in the survey and contained a brief description of the nature of the survey.

The fieldwork was conducted by 75 interviewers from CSU's panel. They were fully briefed and familiarised with the survey procedures. The first briefing session was held on 29th April 1996 with fieldwork beginning immediately afterwards. The main field period extended until 12th July 1996. A small proportion of interviews were carried out in the period between 15th July and 15th August 1996.

The survey was conducted as an SCPR survey, with all survey documents clearly identifying that research organisation. Interviewers however, carried and presented their normal CSU identity cards. To avoid any confusion on the part of respondents, interviewers also carried, and left with respondents, a letter of introduction from the research team at SCPR. The letter clearly identified the relationship between NISRA and SCPR in the context of the survey. Respondents were given the London telephone number of the Social Attitudes research team at SCPR as well as a Belfast telephone number, in case they had any queries or uncertainties about the survey or the interviewer. The Belfast telephone number was a direct telephone line manned by NISRA field-staff during office hours, and otherwise covered by an answering machine. Only a very small number of respondents used either method of contact.

A total of 1400 addresses were selected. They were assigned to interviewers using CSU's normal allocation procedures which ensure minimum travelling distances for each interviewer. The fieldwork was supervised by CSU using the standard quality control methods employed on all government surveys in Northern Ireland. Interviewers were required to make at least three calls at an address (normal procedure allows for additional calls to be made should the interviewer be passing the address while working in the area), before declaring it a non-contact and returning the allocation sheet to headquarters. The timing of the initial contact calls was left to the discretion of the interviewer, based on knowledge of the area, to maximise the likelihood of finding someone at home. Before declaring an address to be a non-contact, at least one call must have been made in the morning, afternoon and evening or weekend.

Table A4   Response rate

	Number	%
Addresses issued	1400	
Vacant, derelict etc.	128	
In scope	1272	100
Interview achieved	786	62
Interview not achieved	486	38
Refused	279	22
Non-contact	151	12
Unproductive interview	56	4

Field-staff at CSU monitored the return of work and quality assured the data which were returned on a weekly basis by the interviewers. Staff at CSU maintained telephone contact with all interviewers and dealt with any problems that arose in the field initially by this means.

An overall response rate of 62 per cent was achieved, based on the total number of issued addresses which were within the scope of the survey (that is, occupied, private addresses). Refusals were obtained at 22 per cent of eligible addresses. At 12 per cent of addresses interviewers could not contact either the household or the selected respondent within the field period. Unproductive interviews were obtained at a further 4 per cent of addresses.

Self-completion Questionnaire

At the end of the face-to-face interview, interviewers introduced the self-completion questionnaire. Where possible, the selected respondent completed the questionnaire while the interviewer was still in the house. If this was not possible, the questionnaire was either collected by the interviewer at a later date, or posted by the respondent to a Northern Ireland Post Office Box. The self-completion questionnaire was then forwarded, through CSU to SCPR.

Figure A2   Summary of response to the survey

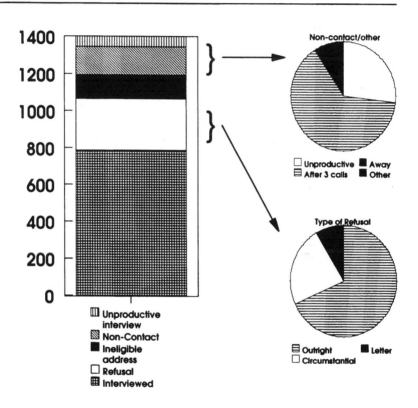

The return of the self-completion questionnaires was monitored by CSU field-staff. Up to two reminder letters were sent at two weekly intervals after the initial interview. In all, 79 per cent of the self-completion questionnaires were returned.

Table A5   Response rate for the self-completion questionnaire

	Number	%
Achieved interviews	786	100
Self-completion returned	620	79
Not returned	166	21

Data Processing and Coding

Disks containing interview information were returned by the field force on a weekly basis. The information contained on the returned disks was transferred onto an office Field Management System on a weekly basis. This procedure meant that the progress of the fieldwork could be monitored on a week by week basis. After the completion of the fieldwork period, final checks were made on the information contained on the return disks prior to the datafiles being sent to SCPR for checking, coding and editing. For the self-completion questionnaire, SCPR conducted all checking, editing, coding, keying and computer editing.

Analysis Variables

The analysis variables in the Northern Ireland dataset are the same as those in the British survey. However, the questions on party identification use Northern Irish political parties. A number of analysis variables were coded by SCPR from the current or last job held by the respondent (and spouse or partner). Summary variables derived from these and some further derived variables are included in the dataset. The principal analysis variables available in the dataset are listed below.

Table A6   Analysis variables

Coded analysis variables	Standard Occupation Classification (1990)
	Employment status
	Socio-economic group (SEG)
	Registrar General's Social Class (I to V)
	Goldthorpe class schema
	Standard Industrial Classification
	(SIC, 1980)
Derived analysis variables	Age within sex
	Current economic position
	Area
	Highest educational qualifications
	Accommodation tenure
	Marital Status

Sampling Errors

No sample is likely to reflect precisely the characteristics of the population it is drawn from, because of both sampling and non-sampling errors. An estimate of the amount of error due to the sampling process can be calculated.

For a simple random sample design, in which every member of the sampled population has an equal and independent chance of inclusion in the sample, the sampling error of any percentage, p, can be calculated by the formula

$$s.e.\ (p) = x\ \sqrt{p(100-p)/n}$$

where n is the number of respondents on which the percentage is based. The sample for the Northern Ireland Social Attitudes survey is drawn as a simple random sample, and thus this formula can be used to calculate the sampling error of any percentage estimate from the survey. A confidence interval for the population percentage can be calculated by the formula

$$95\ per\ cent\ confidence\ interval = p \pm 1.96\ x\ s.e.\ (p)$$

If 100 similar, independent samples were chosen from the same population, 95 of them would be expected to yield an estimate for the percentage, p, within this confidence interval.

The absence of design effects in the Northern Ireland survey, and therefore, of the need to calculate complex standard errors, means that the standard error and confidence intervals for percentage estimates from the survey are only slightly greater than for the British survey, despite the smaller sample size.   It also means that standard statistical tests of significance (which assume random sampling) can be applied directly to the data.

A percentage estimate of 10 per cent (or 90 per cent) which is based on all respondents to the Northern Ireland survey has a standard error of 1.1 per cent and a 95 per cent confidence interval of ± 2.1 per cent. A percentage estimate of 50 per cent has a standard error of 1.8 and a 95 per cent confidence interval of ± 3.5 per cent. Sampling errors for proportions based on sub-groups within the sample are somewhat larger than they would have been had the questions been asked of everyone.

Table A1.1 provides examples of the sampling errors and confidence intervals for typical percentage estimates from the Northern Ireland Social Attitudes survey.

Representativeness of the Sample

In any survey, there is a possibility of non-response bias. Non-response bias arises if the characteristics of non-respondents differ significantly from those of respondents, in such a way that they are reflected in the responses given in the survey. Accurate estimates of non-response bias can only be obtained by comparing characteristics of the achieved sample with the distribution of the same characteristics in the population at the time of sampling.   Such comparisons are usually made to current Census of Population data.

It is not possible to estimate directly whether any non-response bias exists in the Northern Ireland Social Attitudes survey. However, tables (at the end of this appendix) compare the characteristics of both the households and individuals sampled with those sampled in the Continuous Household Survey (CHS) for the 1996-97 year (the survey year running from April 1996 to March 1997). The CHS has a much larger sample (around 3000 households are interviewed) and uses the same simple random sample design. All adults aged 16 or over are interviewed. No weighting is required to compensate for the effect of household size on probability of selection. The CHS has been running for 12 years and produces consistent estimates from year to year.

Where available, figures from the 1991 Census of Population for Northern Ireland have been shown for comparison.

References

Jowell, R., Curtice, J., Park, A., Brook, L., Thomson, K. and Bryson, C. (eds), (1997), *British Social Attitudes: the 14th Report*, Ashgate Publishing Company, Aldershot.

Jowell, R., Brook, L. and Dowds, L. (eds), (1993) *International Social Attitudes: the 10th BSA Report*, Dartmouth, Aldershot.

Sweeney, K. and McClelland, A., (1995), Appendix 1: Technical Details of the Survey in L. Dowds, P. Devine and R. Breen, (eds), *Social Attitudes in Northern Ireland: The Fourth Report,* Appletree Press, Belfast.

The Northern Ireland Census 1991: Summary Report, (1992) HMSO, Belfast.

Table A1.1   Standard errors and confidence limits

	% (p)	Standard error of p (%)	95% confidence interval +/-	95% confidence limits
Classification variables, n=1510				
Derived religion				
Protestant	48.8	1.8	3.5	45.3 - 52.3
Catholic	36.8	1.7	3.4	33.4 - 40.4
Other	14.4	1.2	2.4	12.0 - 16.8
(Tenure 2) Housing Tenture*				
Owns	69.2	1.6	3.2	66.0 - 72.4
Rent from NIHE	24.2	1.5	3.0	21.2 - 27.2
Derived employment status				
Working	54.9	1.8	3.5	51.4 - 58.4
Unemployed	6.4	0.9	1.7	4.7 - 8.1
Attitudinal variables (all), n=1510				
(GPCHANGE) consider it not difficult to change GP	67.6	1.7	3.3	64.3 - 70.9
Version B, n=744				
(ECGBCLSE) The UK should have closer links with the European Community	35.8	1.9	3.8	32.0 - 39.6
Employees only, n=627				
(INDREL) Not good relations between management and employees	14.1	1.8	3.6	10.5 - 17.7

*Unweighted data

Table A1.2 Comparison of household characteristics

		NISA Survey 1996*	Continuous Household Survey 1996/7	Northern Ireland Census 1991
Characteristics of sampled households				
Tenure	Owner occupied	69	67	62
	Rented, NIHE	24	25	29
	Rented, Other	7	6	8
	Rent free	1	1	1
Type of home	Detached	35	33	31
	Semi-detached	26	24	23
	Terraced	32	33	37
	Purpose-built flat	5	6	7
	Converted flat	1	1	2
	Other	-	2	-
Household income (£)	Less than 4,000	9	9	
	4,000 - 7,999	24	24	
	8,000 - 11,999	12	13	
	12,000 - 17,999	12	14	
	18,000 - 19,999	4	5	
	20,000 and over	25	24	
	Unknown	13	11	
Base = 100%		786	2,812	530,369

*Household characteristics are based on unweighted data from the NISA survey

Table A1.3 Comparison of individual characteristics

		NISA*** Survey 1996	Continuous Household Survey 1996/97	Northern Ireland Census 1991
Individual Characteristics				
Sex	Male	45	47	48
	Female	55	52	52
Age	18 - 24	10	12	16
	25 - 34	23	21	21
	35 - 44	20	19	18
	45 - 54	17	16	15
	55 - 59	7	7	6
	60 - 64	6	6	6
	65 and over	16	19	18
Marital Status	Single	24	26	28
	Married/ Cohabiting	62	60	59
	Widowed	8	8	9
	Divorced/ separated	5	6	6
Economic activity	Working	55	51	49*
	Unemployed	6	6	9
	Activity	39	39	42
	Refused/ missing	-	4	-
Base = 100 %		786	5,586	1,117,221 **

---

*   Based on total population aged 16 and over (base = 1,167,938).
**   Persons aged 18 and over
***   Participating selected respondents

Table A1.4 Stated religious denomination (%)

Religious denomination of persons aged 18 years and over	NISA Survey 1996	Continuous Household Survey 1996/97	Northern Ireland Census 1991**
Protestant	49	59	50
Catholic	37	38	38
Non-Christian	-	-	-
No religion	14	3	4
Unwilling to say	-	1	7
Base = 100%	786	4650	1,577,836
(Undefined CHS*)		(17%)	

*   Religion remains undefined in the CHS for individuals who did not fully co-operate in the survey and were, therefore not asked their denomination. The base for this percentage (5572) is the total number of adults aged 18 and over in the sampled households.
**   Usually resident population (all ages)

Table A1.5 Redefined religious denomination (%)

Religious denomination* of persons aged 18 years and over	NISA Survey 1996	Continuous Household Survey 1996/97
Protestant	55	57
Catholic	42	39
Non-Christian	0	-
No religion	1	3
Unwilling to say	1	1
Base = 100%	786	5,572

*   Religious denomination has been redefined in both surveys, for those who stated 'No religion' or were unwilling to specify their denomination. In the NISA survey, denomination was calculated from the religion in which the respondent was brought up. In the CHS, denomination was redefined using the denomination specified by other members of the household.

Appendix II:
Notes on the Tabulations

1. Figures in the tables are from the 1996 Northern Ireland Social Attitudes survey unless otherwise indicated.
2. Tables are percentaged as indicated.
3. In tables, '*' indicates less than 0.5 per cent but greater than zero, and '-' indicates zero.
4. When findings based on the responses of fewer than 100 respondents are reported in the text, reference is generally made to the small base size.
5. Percentages equal to or greater than 0.5 have been rounded up in all tables (e.g. 0.5 per cent = 1 per cent, 36.5 per cent = 37 per cent).
6. In many tables the proportions of respondents answering 'Don't know' or not giving an answer are omitted. This, together with the effects of rounding and weighting, means that percentages will not always add to 100 per cent.
7. The self-completion questionnaire was not completed by all respondents to the main questionnaire (see Appendix I). Percentage responses to the self-completion questionnaire are based on all those who completed it.

Appendix III:
Using Northern Ireland Social Attitudes Survey Data

All survey datasets are deposited and can be obtained from the ESRC Data Archive at Essex University. In addition, the survey years 1989-1991 are available as a fully documented *combined* dataset. Although the annual book covers many topics in depth, it cannot hope to provide time-series data for all questions included in that survey round; for that reason we would encourage interested parties to use the data directly.

The core of the survey is the community relations module which has been included in every survey round except for 1990 when a module on attitudes to crime and the police was fielded instead. The list on the following page shows the modules fielded in every survey year.

Topics covered* in Northern Ireland Social Attitudes surveys 1989-1995

Topics (excluding 'core' ones)	Survey Year						
	1989	1990	1991	1993	1994	1995	1996
AIDS	✓						✓
Attitudes to work (ISSP)	✓						
Changing gender roles (ISSP)					✓		
Charitable giving			✓	✓			
Childcare					✓	✓	
Civil liberties		✓			✓		
Community relations	✓		✓	✓	✓	✓	✓
Countryside and the environment		✓		✓	✓	✓	✓
Crime and the police			✓				
Diet and health	✓						
Drugs							✓
Economic prospects	✓	✓	✓	✓		✓	✓
Education				✓		✓	
Family networks							✓
Gender issues at the workplace			✓		✓		
Gender roles			✓		✓		
Global environmental issues (ISSP)				✓			
Health and lifestyle			✓				
Housing		✓					✓
Informal carers						✓	
National identity (ISSP)							✓
National Health Service	✓	✓	✓	✓	✓	✓	✓
Police and the public		✓					
Political trust					✓		✓
Poverty	✓					✓	
Race and immigration						✓	
Religious beliefs (ISSP)			✓				
Role of government (ISSP)		✓					✓
Single parenthood and child support				✓	✓	✓	
Sexual morality	✓	✓					
Social class	✓	✓			✓		
Taxation and public spending						✓	✓
Transport						✓	
UK's relations with Europe/other countries	✓	✓	✓	✓	✓	✓	✓
Welfare state				✓		✓	✓

*   Excluded are 'core topics' such as public spending, workplace issues and economic prospects, and standard classification items such as economic activity, newspaper readership, religious denomination and party identification, all of which are asked every year.

Appendix IV:
The Questionnaires

As explained in Appendix I the NISA survey was carried out using Computer Assisted Interviewing (CAI). The questionnaire reproduced here was derived from the Blaise program in which it was written. The keying codes have been removed and the percentage distribution of answers to each question inserted instead. The SPSS variable name is also included. Routing directions are given above each question and any routing instruction should be considered as staying in force until the next routing instruction.

Percentages are based on the total weighed sample, 786 main questionnaire and 620 self-completion questionnaire. Some questions are filtered - that is, they are asked of only a proportion of respondents. In these cases the weighted base is indicated at the beginning of that question. The percentage distributions do not necessarily add up to 100 because of weighting and rounding, or because at a few questions, respondents were invited to give more than one answer and so percentages may add to well over 100 per cent. These are clearly marked by interviewer instructions on the questionnaires.

For further details on the questionnaires readers are referred to Jowell, *et al* (1997).

References

Jowell, R., Curtice, J., Park, A., Brook, L., Thompson, K. and Bryson, C. (1997) *British Social Attitudes: The 14th Report*, Ashgate, Aldershot.

NORTHERN IRELAND SOCIAL ATTITUDES: 1996 FACE-TO-FACE INTERVIEW

DOCUMENTATION

Contents page

NEWSPAPER READERSHIP/POLITICS

ASK ALL
Q215 [Readpap]   n=786
Do you normally read any daily **morning** newspaper at least 3 times a week?
%
55.0   Yes
45.0   No
-   (Don't know)

IF 'YES' AT [ReadPap]
Q216 [WhPaper]   n=433
Which one do you normally read?
IF MORE THAN ONE: Which one do you read **most** frequently?
%
3.9   (Scottish) Daily Express
4.0   (Scottish) Daily Mail
18.9   Daily Mirror/Record
3.4   Daily Star
26.7   The Sun
2.6   Daily Telegraph
0.1   Financial Times
1.0   The Guardian
1.3   The Independent
4.1   The Times
-   Morning Star
15.8   The News Letter
15.3   The Irish News
1.0   The Irish Times
0.4   Other Irish/Northern Irish/Scottish regional or local **daily** morning paper **(WRITE IN)**
1.3   Other **(WRITE IN)**
-   (Don't Know)

ASK ALL
Q219 [Politics]   n=786
How much interest do you generally have in what is going on in politics ... **READ OUT** ...
%
7.7   ... a great deal,
20.0   quite a lot,
31.7   some
29.0   not very much
11.6   or, none at all?
-   (Don't know)

PUBLIC SPENDING, WELFARE BENEFITS AND HEALTH CARE

ASK ALL
Q221 [Spend1]   n=786
CARD
Here are some items of government spending. Which of them, if any, would be your highest priority for extra spending?
Please read through the whole list before deciding.
ENTER ONE CODE ONLY FOR HIGHEST PRIORITY

IF ANSWER GIVEN AT [Spend1] (I.E. NOT 'None of these/DK/Refusal')

Q222 [Spend2]

CARD AGAIN
And which next?
ENTER ONE CODE ONLY FOR NEXT HIGHEST

	[Spend1] %	[Spend2] %
Education	21.3	41.7
Defence	0.2	0.5
Health	62.6	24.8
Housing	4.1	7.8
Public transport	0.9	2.1
Roads	2.0	3.9
Police and prisons	0.5	0.9
Social security benefits	4.4	9.8
Help for industry	3.1	7.4
Overseas aid	0.5	0.6
(None of these)	0.1	0.1
(Don't Know)	0.2	0.1
(Refusal/NA)	-	0.3

ASK ALL
Q223 [SocBen1]
CARD
Thinking now only of the government's spending on **social benefits** like those on the card.
Which, if any, of these would be your highest priority for **extra** spending?
ENTER ONE CODE ONLY FOR HIGHEST PRIORITY

IF ANSWER GIVEN AT [SocBen1]
(I.E. NOT 'None of these/DK/Refusal;)
Q224 [SocBen2]   n=786
CARD AGAIN
And which next?
ENTER ONE CODE ONLY FOR NEXT HIGHEST

	[SocBen1]	[SocBen2]
	%	%
Retirement pensions	42.3	22.9
Child benefits	17.0	18.4
Benefits for the unemployed	10.4	10.2
Benefits for disabled people	23.0	37.2
Benefits for single parents	6.7	10.0
(None of these)	0.4	1.0
(Don't Know)	0.2	0.2
(Refusal/NA)	-	0.2

ASK ALL
Q225 [Dole]
Opinions differ about the level of benefits for unemployed people.
Which of these two statements comes closest to your own view
... READ OUT ...

%
49.3 ... benefits for unemployed people are too low and cause hardship,
31.4 or, benefits for unemployed people are too high and discourage them from finding jobs?
13.5 (Neither)
0.3   Both low wages
0.6   Both varies
0.2   About right
2.1   Other Answer (WRITE IN)
2.5   (Don't Know)

Q227 [TaxSpend]
CARD
Suppose the government had to choose between the three options on this card. Which do you think it should choose?

%
4.4   Reduce Taxes and spend less on health, education and social benefits
36.7 Keep taxes and spending on these services at the same level as now
55.2 Increase taxes and spend more on health, education and social benefits
2.8   (None)
0.9   (Don't Know)

3

Q228 [NHSSat]   n=786
CARD
All in all, how satisfied or dissatisfied would you say you are with the way in which the National Health Service runs nowadays?
Choose a phrase from this card.

Q229 [GPSat]
CARD AGAIN
From your own experience or from what you have heard, please say how satisfied or dissatisfied you are with the way in which each of these parts of the National Health Service runs nowadays:
First, local doctors or GPs?

Q230 [DentSat]
CARD AGAIN
(And how satisfied or dissatisfied are you with the NHS as regards ...)
... National Health Service dentists?

	[NHSSat]	[GPSat]	[DentSat]
	%	%	%
Very satisfied	6.1	34.8	23.9
Quite satisfied	31.0	47.9	47.3
Neither satisfied nor dissatisfied	15.6	5.2	12.8
Quite dissatisfied	24.3	9.5	9.0
Very dissatisfied	22.5	2.2	4.4
(Don't Know)	0.4	0.3	2.6

Q231 [InpatSat]
CARD AGAIN
(And how satisfied or dissatisfied are you with the NHS as regards ...)
... Being in hospital as in-patient?

4

Q232 [OutpaSat]
CARD AGAIN   n=786
(And how satisfied or dissatisfied are you with the NHS as regards ...)
... Attending hospital as an out-patient?

	InpatSat	OutpaSat
	%	%
Very Satisfied	23.2	15.0
Quite Satisfied	38.8	14.8
Neither Satisfied nor dissatisfied	15.5	13.4
Quite Dissatisfied	13.1	15.7
Very dissatisfied	5.4	6.8
(Don't Know)	4.1	4.3

Q233 [PrivMed]
Are you yourself covered by a private health insurance scheme, that is an insurance that allows you to get private medical treatment?
ADD IF NECESSARY: 'For example, BUPA or PPP.'
IF INSURANCE COVERS DENTISTRY <u>ONLY</u>, **CODE 'No'**
%
8.2 Yes
91.8 No
- Don't Know

IF 'Yes' AT [PrivMed]
Q234 [PrivPaid]
Does your employer (or your partner's employer) pay the majority of the cost of membership of this scheme?
%
26.4 Yes
73.6 No
- Don't Know

ASK ALL
Q235 [NHS Limit]
It has been suggested that the National Health Service should be available **only to those with lower incomes.**
This would mean that contributions and taxes could be lower and most people would then take out medical insurance or pay for health care.
Do you support or oppose this idea?
%
8.3 Support a lot
16.7 Support a little
16.3 Oppose a little
56.1 Oppose a lot
2.3 (Don't Know)
0.2 (Refusal/NA)

5

Q236 [WhchHosp]
CARD   n=786
Now suppose you needed to go into hospital for an operation.
Do you think you would have a say about which hospital you went to?
%
14.1 Definitely would
25.8 Probably would
40.1 Probably would not
16.3 Definitely would not
3.4 (Don't Know)

Q237 [GPChange]
Suppose you wanted to change your GP and go to a different practice, how difficult or easy do you think this would be to arrange?
Would it be ... **READ OUT** ....
%
7.7 ... very difficult,
14.9 Fairly difficult,
36.2 not very difficult,
31.4 or, not at all difficult?
9.7 (Don't Know)

Q238 [DenLimNI]
Some dentists now provide NHS treatment only to those with lower incomes.
This means that other people have to pay the full amount for their dental treatment, or take out private insurance to cover their treatment.
Do you support or oppose this happening?
IF 'SUPPORT' OR 'OPPOSE' : A lot or a little?
%
6.3 Support a lot
17.7 Support a little
21.9 Oppose a little
52.2 Oppose a lot
1.8 (Don't Know)

IF 'YES' AT [PrivMed]
Q239 [DentInsu]   n=64
Does the private medical insurance scheme you belong to cover your treatment at the dentist?
%
16.9 Yes
74.3 No
1.3 (Don't go to the dentist)
7.4 (Don't Know)

6

IF 'No' AT [PrivMed] OR AT [DentInsu]
Q240 [DentOthr]   **n=775**
Is your dental treatment covered by any (other) Private insurance scheme?
%
1.6 Yes
95.9 No
1.9 (Don't go to the dentist)
- (Don't Know)
0.6 (Refusal/NA)

ASK ALL
Q242 [FagsNow2]   **n=786**
% Do you yourself ever smoke cigarettes?
32.5 Yes
67.5 No
- (Don't Know)

IF 'YES' AT [Cigs]
Q243 [SmokDay]   **n=255**
About how many cigarettes a day do you usually smoke?
IF CAN'T SAY, CODE 997
Range: 0... 997
Median: 15.0

ASK ALL
Q246 [ReconAct]   **n=786**
CARD
Which of these descriptions applies to what you were doing last week, that is, in the seven days ending last Sunday?
PRIORITY CODING - FIRST CODE THAT APPLIES
%
3.6 In full-time education (not paid for by employer, including on vacation)
0.8 On government training/employment programme (eg. Youth Training, Training for Work etc)
54.9 In paid work (or away temporarily) for at least 10 hours in week
0.1 Waiting to take up paid work already accepted
5.6 Unemployed and registered at a benefit office
0.3 Unemployed, not registered, but actively looking for a job (or at least 10 hours a week)
0.5 Unemployed, wanting a job (of at least 10 hours per week) but not actively looking for a job
6.2 Permanently sick or disabled
14.3 Wholly retired from work
13.5 Looking after the home
0.3 (Doing something else) (WRITE IN)
- (Don't Know)

7

IF 'In full-time education', 'On government training', 'Unemployed', 'Wholly retired', 'Looking after the home' OR 'Doing something else' AT (REconAct) (I.E. NOT WORKING)
Q247 [RLastJob]   **n=354**
How long ago did you last have a paid job of at least 10 hours a week?
GOVERNMENT PROGRAMS/SCHEMES DO NOT COUNT AS 'PAID JOBS'.
%
13.0 Within past 12 months
23.1 Over 1, up to 5 years ago
17.4 Over 5, up to 10 years ago
18.3 Over 10, up to 20 years ago
16.1 Over 20 years ago
12.1 Never had a paid job of 10+ hours a week
- (Don't Know)

ASK ALL WHO HAVE EVER WORKED (IF 'In paid work' OR 'Waiting to take up paid work'' AT [REconAct] OR EVER HAD A PAID JOB AT [RLastJob])
Q248 [Rtitle] **(NOT ON THE DATA FILE)**
IF IN PAID WORK (IF 'In paid work' AT [REconAct]) : Now I want to ask you about your present job. What is your job?
PROBE IF NECESSARY: What is the name or title of the job?
IF WAITING TO TAKE UP PAID WORK (IF 'Waiting to take up paid work' AT [REconAct]): Now I want to ask you about your future job. What is your job?
PROBE IF NECESSARY: What is the name or title of the job?
IF NOT IN PAID WORK (OR WAITING TO TAKE UP PAID WORK) BUT EVER HAD JOB IN THE PAST (AT [LastJob]): Now I want to ask you about your last job. What was your job?
PROBE IF NECESSARY: What was the name or title of the job?
Open question (Maximum of 50 characters)

Q249 [RTypeWK] **(NOT ON THE DATA FILE)**
What kind of work (do/will/did) you do most of the time?
IF RELEVANT: What materials/machinery (do/will/did) you use?
Open question (maximum of 50 characters)

Q250 [Rtrain] **(NOT ON THE DATA FILE)**
What training or qualifications (are/were) needed for that job?
Open question (Maximum of 50 characters)

8

Q251 [Rsuper2] n=743
(Do/will/did) you directly supervise or (are you/will you be/were you) directly responsible for the work of any other people?
%
35.4 Yes
64.6 No
- (Don't Know)

IF 'Yes' AT [Rsuper2] n=262
Q252 [Rmany]
How many?
Range: 0 ... 9997
Median: 5.0

Q253 [Rsuper]
Derived from [Rsuper2]

ASK ALL WHO HAVE EVER WORKED (IF 'In paid work' OR 'Waiting to take up paid work' AT [REconAct] OR EVER HAD A PAID JOB AT [RLastJob])

Q254 [RSupMan]
% Can I just check, (are you/will you be/were you) ... READ OUT ...
15.3 ... a manager,
16.9 a foreman or supervisor,
67.8 or not?
- (Don't Know)

Q255 [Remplyee]
% In your (main) job (are you/will you be/were you ... READ OUT ...
89.1 ... an employee,
10.9 or self-empoyed?
- (Don't Know)

ASK IF 'In paid work' AT [REconAct]
Q256 [Remplye]
% Derived from [REmplyee]
85.7 Employed
14.3 Self-employed
- (Don't Know)

ASK IF 'employee/DK' AT [REmployee] n=662
Q257 [ROcSect]
CARD
Which of the types of organisation on this card (do you work/will you be working/did you work) for?
%
59.5 Private sector firm or company (including limited companies and plcs)
1.6 Nationalised industry/public corporation
10.7 Local authority/Education and Library Board
13.6 Health authority/NHS hospital/NHS Hospital Trust (including GP surgeries)
10.7 Central government/Civil service/Government Agency
Charity/Voluntary sector (including charitable companies)
1.6 Other answer (WRITE IN)
0.1 (Don't Know)
0.1 (Refusal/NA)

ASK ALL WHO HAVE EVER WORKED (IF 'In paid work' OR 'Waiting to take up paid work' AT [REconAct] OR EVER HAD A PAID JOB AT [RLastJob])

Q259 [REmpMake] (NOT ON THE DATA FILE)
What (does/did) your employer (IF SELF-EMPLOYED: you) make or do at the place where you (work/will work/worked) (from)?
Open Question (Maximum of 80 characters)

ASK IF 'self employed' AT [REmplyee] n=62
Q260 [SPartnr]
In your work or business, do you have any partners or other self-employed colleagues?
%
52.2 Yes, has partner (s)
47.8 No
- (Don't Know)

ASK IF 'self-employed' AT [REmplyee]
Q262 [SEmpNum]
In your work or business, (do/did) you have any employee, or not?
IF YES: How many?
IF 'NO EMPLOYEES' CODE 0.
FOR 500+ EMPLOYEES, CODE 500
NOTE: FAMILY MEMBERS MAY BE EMPLOYEES ONLY IF THEY RECEIVE A REGULAR WAGE OR SALARY.
Range: 0 500
Median: 1.0

ASK ALL WHO HAVE EVER WORKED (IF 'In paid work' OR 'Waiting to take up paid work' AT [REconAct] OR EVER HAD A PAID JOB AT [RLastJob])   n=724

Q263   [REmpWork]
Including yourself, how many people (are/were) employed at the place where you usually (work/will work/worked) (from)?

%
4.1   None
20.4   Under 10
16.0   10-24
25.0   25-99
19.6   100-499
13.2   500 or more
1.6   (Don't Know)
0.1   (Refusal/NA)

ASK IF 'self-employed' AT [REmploye]   n=62

Q265   [SNumEmp]
Derived variable from [REmploye] and [SEmpNum]
%
56.7   Yes
43.3   No
-    (Don't Know)

ASK IF 'In paid work' AT [REconAct]   n=431

Q266   [RWKJbTim]
In your present job, are you working
... READ OUT ...
RESPONDENT'S OWN DEFINITION
%
80.4   ... full-time,
19.6   or, part-time?
-    (Don't Know)

Q 269   [WkJbHrsl]
How many hours do you normally work a week in your main job - including any paid or unpaid overtime?
ROUND TO NEAREST HOUR.
IF RESPONDENT CANNOT ANSWER, ASK ABOUT LAST WEEK.
IF RESPONDENT DOES NOT KNOW EXACTLY, ACCEPT AN ESTIMATE.
FOR 95+ HOURS, CODE 95.
FOR 'VARIES TOO MUCH TO SAY', CODE 96.
Range: 10 ... 96
Median: 39.0

ASK IF 'employed/DK' AT [REmploye]   n=369

Q270   [EJbHrsX]

What are your basic or contractual hours each week in your main job - excluding any paid and unpaid overtime?
ROUND TO NEAREST HOUR.
IF RESPONDENT DOES NOT KNOW EXACTLY, ACCEPT AN ESTIMATE.
FOR 95+ HOURS, CODE 95.
FOR 'VARIES TOO MUCH TO SAY', CODE 96.
Range: 0 ... 96
Median: 37.0

ASK IF 'waiting to take up paid work' AT [REconAct] OR EVER HAD A PAID JOB AT [RLastJob]

Q271   [ExPrtFul]
%   (Is/was) the job ... READ OUT ...
73.2   ... full-time - that is, 30 or more hours per week,
26.8   or, part-time?
-    (Don't Know)

ASK ALL WHO HAVE EVER WORKED (IF 'In paid work' OR 'Waiting to take up paid work' AT [REconAct] OR EVER HAD A PAID JOB AT [RLastJob])   n=743

Q301   [UnionSA]
(May I just check) are you now a member of a trade union or staff association?
CODE FIRST TO APPLY
%
23.4   Yes, trade union
4.4   Yes, staff association
72.2   No
-    (Don't Know)

IF 'No/DK' AT [UnionSA]   n=537

Q302   [Tusaever]
Have you ever been a member of a trade union of staff association?
CODE FIRST TO APPLY
%
28.6   Yes, trade union
2.5   Yes, staff association
68.9   No
0.1   (Don't Know)

ASK IF NOT IN PAID WORK AT [REconAct]   **n=355**

Q306 [NPWork10]
In the seven days ending last Sunday, did you have any paid work of less than 10 hours a week?
%
2.3   Yes
97.7   No
-      (Don't Know)

ASK IF 'employee/DK' AT [REmploye]   **n=369**

Q307 [WageNow]
How would you describe the wages or salary you are paid for the job you do - on the low side, reasonable, or on the high side?
        IF LOW: Very low or a bit low?
%
8.7   Very low
25.2   A bit low
61.7   Reasonable
3.7   On the high side
0.7   Other answer (WRITE IN)
-      (Don't Know)

Q309 [PayGap]
CARD
Thinking of the highest and the lowest paid people in your place of work, how would you describe the gap between their pay, as far as you know?
        Please choose a phrase from this card.
%
19.8   Much too big a gap
25.7   Too big
42.9   About right
4.5   Too small
0.2   Much too small a gap
6.9   (Don't Know)

Q310 [WageXpct]
If you stay in this job, would you expect your wages or salary over the coming year to ... READ OUT ...
%
14.7   ... rise by more than the cost of living,
47.7   rise by the same as the cost of living,
22.5   rise by less than the cost of living
13.2   or, not to rise at all?
1.2   (Will not stay in job)
0.6   (Don't Know)

IF 'not rise at all' AT [WageXpct]   **n=51**

Q311 [WageDown]
Would you expect your wages or salary to stay the same, or in fact to go down?
%
92.8   Stay the same
3.0   Go down
4.2   (Don't Know)

ASK IF 'employee/DK' AT [REmploye]   **n=369**

Q312 [NumEmp]
Over the coming year do you expect your workplace to be ... READ OUT ...
%
20.3   ... increasing its number of employees,
22.5   reducing its number of employees,
55.7   or, will the number of employees stay about the same?
1.5   Other answer (WRITE IN)
-      (Don't Know)

Q314 [LeaveJob]
Thinking now about your own job.
How likely or unlikely is it that you will leave this employer over the next year for any reason?
        Is it ... READ OUT ...
%
10.1   ... very likely,
10.1   quite likely,
25.0   not very likely,
53.1   or, not at all likely?
1.8   (Don't Know)

13

14

IF 'very likely' OR 'quite likely' AT [LeaveJob] n=81
Q315 [WhyGo8]
Why do you think you will leave? Please choose a phrase from this card or tell me what other reason there is.

	Yes %	No %	DK %
Firm will close down	3.2	88.8	8.0
I will be declared redundant	1.7	90.3	8.0
I will reach normal retirement age	3.1	88.9	8.0
My contract of employment will expire	10.1	81.9	8.0
I will take early retirement	7.2	84.8	8.0
I will decide to leave and work for another employer	51.8	40.2	8.0
I will decide to leave work for myself, as self-employed	5.4	86.6	8.0
I will look after home/children/relative	3.0	89.0	8.0
Return to education	4.4	87.6	8.0
Other Answer	8.8	83.2	8.0
(Don't Know) (WRITE IN)	-	-	-

ASK IF 'employee/DK' AT [REmploye] n=369
Q327 [ELookJob]
Suppose you lost your job for one reason or another - would you start looking for another job, would you wait for several months or longer before you started looking, or would you decide not to look for another job?
%
83.3 Start looking
8.7 Wait several months or longer
7.6 Decide not to look
0.4 (Don't Know)

IF 'start looking' AT [REmploye] n=309
Q330 [EFindJob]
How long do you think it would take to find an acceptable replacement job?
Range: 1 ... 997
Median: 2.0

IF 3 MONTHS OR MORE OR DK AT [EfindJob] n=163
Q331 [Eretrain]
How willing do you think you would be in these circumstances to retrain for a different job ... READ OUT ...
%
55.7 ... very willing,
24.5 quite willing,
19.0 or - not very willing?
- (Don't Know)
0.8 (Refusal/NA)

IF 'employee/DK' AT [REmploye] n=369
Q332 [ESelfEm]
For any period during the last five years, have you worked as a self-employed person as your main job?
%
3.2 Yes
96.8 No
- (Don't Know)

ASK IF NOT EMPLOYED AT [REconAct] n=573
Q333 [NwUnemp]
During the last five years - that is since May 1991 - have you been unemployed and seeking work for any period?
%
15.0 Yes
85.0 No
- (Don't Know)

IF 'Yes' AT [NwUnemp]
Q334 [NwUnempT]
For how many months in total during the last five years (that is since May 1991 have you been unemployed and seeking work?)
INTERVIEWER: IF LESS THAN ONE MONTH, CODE AS 1.
Range: 1 ... 60
Median: 10.0

ASK IF UNEMPLOYED AT [REconAct] n=50
Q339 [JobQual]
How confident are you that you will find a job to match your qualifications ... READ OUT ...
%
18.5 ... very confident,
27.8 quite confident,
25.8 not very confident,
27.9 or, not at all confident?
- (Don't Know)

Q342 [UFindJob]   n=50
Although it may be difficult to judge, how long from now do you think it will be before you find an acceptable job?
Range: 1 ...997
Median: 6.0 months

IF MORE THAN 2 YEARS AT [UFindJob]
Q343 [URetrain]
How willing do you think you would be in these circumstances to retrain for a different job ... READ OUT ...

Q344 [UJobMove]
How willing would you be to move to a different area to find an acceptable job ... READ OUT ...

Q345 [UBadJob]   n=35
And how willing do you think you would be in these circumstances to take what you now consider to be an unacceptable job ... READ OUT ...

	[URetrain]	[UJobMove]	UBadJob
	%	%	%
... very willing,	19.0	1.5	2.9
quite willing,	33.1	19.1	42.7
or, not very willing?	43.6	79.5	54.4
(Don't Know)	4.3	-	-

ASK IF UNEMPLOYED AT [REconAct]
Q346 [ConMove]   n=50
Have you ever actually considered moving to a different area - an area other than the one you live in now - to try to find work?
%
30.4 Yes
69.6 No
- (Don't Know)

Q347 [UJobChnc]
Do you think that there is a real chance nowadays that you will get a job in this area, or is there no real chance nowadays?
%
47.1 Real chance
50.2 No real chance
2.8 (Don't Know)

17

Q348 [FPtWork]   n=50
% Would you prefer full- or part-time work, if you had the choice?
79.2 Full-time
16.0 Part-time
3.8 Not looking for work
1.0 (Don't Know)

IF 'Part-time' AT [FPtWork]
Q349 [PartTime]   n=8
About how many hours per week would you like to work?
PROBE FOR BEST ESTIMATE
Range: 1 30
Median: 18.9

ASK IF 'Looking after the home' AT [REconAct]
Q350 [EverJob]   n=106
Have you, during the last five years, ever had a full- or part-time job of 10 hours or more a week?
%
16.6 Yes
83.4 No
- (Don't Know)

IF 'No' AT [EverJob]
Q351 [FUJobSer]   n=89
How seriously in the past five years have you considered getting a full-time job?
PROMPT, IF NECESSARY: Full-time is 30 or more hours a week ... READ OUT ...

IF 'not very seriously' OR 'not at all seriously' or DK AT [FtJobSer]
Q352 [PtJobSer]
How seriously, in the past five years, have you considered getting a part-time job ... READ OUT ...

	[FtJobSer]	[PtJobSer]
	%	%
... very seriously,	4.2	3.3
quite seriously,	5.1	8.0
not very seriously,	12.5	14.2
or, not at all seriously?	78.2	74.5
(Don't Know)	-	-

18

ASK IF 'self-employed' AT [REmploye]   n=62
Q353 [Semple]
Have you, for any period in the last five years, worked as an
employee as your main job rather than self-employed?
%
26.7 Yes
73.3 No
-     (Don't Know)

IF 'Yes' AT [Semple]   n=16
Q354 [SempleT]
In total for how many months during the last five years have you been
an employee?
ENTER NUMBER OF MONTHS
Range: 1....60
Median: 48.0

IF 'No/DK' AT [Semple]   n=45
Q355 [SemplSer]
How seriously in the last five years have you considered getting a job
as an employee
... READ OUT ...
%
3.1   ... very seriously,
3.1   quite seriously,
14.2 not very seriously,
79.7 or, not at all seriously?
-     (Don't Know)

ASK IF 'self-employed' AT [REmploye]   n=62
Q356 [Bus1OK]
Compared with a year ago, would you say your business is doing
... READ OUT ...
%
8.0   ... very well,
27.0 quite well,
37.7 about the same,
13.9 not very well,
4.7   or, not at all well?
8.8   (Business not in existence then)
-     (Don't know)

Q357 [Bus1Fut]
And over the coming year, do you think your business will do
... READ OUT ...
34.3 ... better,
41.7 about the same,
19.4 or, worse than this year?
4.5   Other answer (WRITE IN)
-     (Don't Know)

19

ASK IF 'employed' AT [REmploye]   n=369
Q359 [WpUnions]
At your place of work are there unions, staff associations, or groups
of unions recognised by the management for negotiating pay and
conditions of employment?
IF YES, PROBE FOR UNION OR STAFF ASSOCIATION
IF 'BOTH', CODE '1'
%
53.5 Yes : trade union(s)
5.1   Yes : staff association
37.7 No, none
3.7   (Don't Know)

IF 'Yes' : trade unions/staff association   AT [WpUnions]   n=230
Q360 [WpUnsure]
Can I just check: does management recognise these unions or staff
associations for the purposes of negotiating pay and conditions of
employment?
%
87.2 Yes
5.6   No
1.2   (Don't Know)
5.9   (Refusal/NA)

Q361 [WpUnionW]
On the whole, do you think (these unions do their/this staff
association does its) job well or not?
%
54.5 Yes
33.4 No
6.2   (Don't Know)
5.9   (Refusal/NA)

Q362 [TUShould]
CARD
Listed on the card are a number of things trade unions or staff
associations can do. Which, if any, do you think is the most
important thing they should try to do at your workplace?
UNIONS OR STAFF ASSOCIATIONS SHOULD TRY TO:
%
17.4 Improve working conditions
16.1 Improve pay
38.7 Protect existing jobs
5.4   Have more say over how work is done day-to-day
8.2   Have more say over management's long-term plans
3.0   Work for equal opportunities for women
3.6   Reduce pay differences at the workplace
1.0   Work for equal opportunities for ethnic minorities
5.9   Blank
0.2   (None of these)
0.5   (Don't know)

20

ASK IF 'employed'/DK AT [REmploye]   n=369
Q363 [IndRel]

In general how would you describe relations between management and other employees at your workplace ... READ OUT ...
%
34.4 ... very good,
51.1 quite good,
12.0 not very good,
2.1   nor, not at all good?
0.4   (Don't Know)

Q364 [WorkRun]
And in general, would you say your workplace was
%   ... READ OUT ...
31.4 ... very well managed,
53.8 quite well managed,
14.3 or, not well managed?
0.5   (Don't Know)

ASK ALL EXCEPT THOSE WHO ARE 'Wholly retired' OR 'Permanently sick or disabled' AT   [REconAct]   n=625
Q365 [NwEmpEm]
IF IN PAID WORK: Now for some more general questions about your work. For some people their job is simply something they do in order to earn a living. For others it is much more than that. On balance, is your present job ... READ OUT ...
IF NOT IN PAID WORK: For some people work is simply something they do in order to earn a living. For others it means much more than that. In general, do you think of work as
%   ... READ OUT ...
31.3 ... just a means of earning a living,
67.9 or, does it mean much more to you than that?
0.7   (Don't Know)
0.1   (Refusal/NA)

IF 'just a means of earning a living' AT [NwEmpEm]   n=206
Q366 [NwEmpLiv]
Is that because ... READ OUT ...
%
24.7 ... there are no (good) jobs around here,
22.7 you don't have the right skills to get a (good) job
46.0 or, because you would feel the same about any job you had?
3.5   (Don't Know)
3.2   (Refusal/NA)

21

ASK IF IN PAID WORK AT [REconAct]   n=321
Q367 [WkWork1 - WkWork9]
CARD
Now I'd like you to look at the statements on the card and tell me which ones best describe your own reasons for working at present.
PROBE: Which others?
Multicoded (Maximum of 9 codes)

IF MORE THAN ONE ANSWER GIVEN AT [WkWork1-9]
Q379 [WkWkMain]
CARD AGAIN
And which one of these would you say is your main reason for working?

	All reasons [WkWork1-9]	[WkWkMain]
Working is the normal thing to do	28.1	5.1
Need money for basic essentials such as food, rent or mortgage	59.4	52.1
To earn money to buy extras	31.3	10.5
To earn money of my own	21.5	7.1
For the company of other people	25.6	1.8
I enjoy working	47.2	14.2
To follow my career	25.9	6.9
For a change from my children or housework	8.0	0.8
Other answer (WRITE IN)	4.1	1.7
(Don't Know)	-	-
(Refusal/NA)	-	0.3

ASK IF 'employee'/DK AT [REmploye]   n=369
Q380 [SayJob]
Suppose there was going to be some decision made at your place of work that changed the way you do your job. Do you think that you personally would have any say in the decision about the change, or not?
IF 'DEPENDS': Code as 'Don't Know'
%
53.1 Yes
45.2 No
1.7   (Don't Know)

22

IF 'Yes' AT [SayJob]
Q381 [MuchSay]   n=202
How much say or chance to influence the % decision do you think
you would have ... **READ OUT** ...
%
23.1 ... a great deal,
43.2 quite a lot,
30.6 or, just a little?
  -   (Don't Know)
3.1   (Refusal/NA)

ASK IF 'employee'/DK AT [REmploye]   n=369
Q382 [MoreSay]
Do you think you should have more say in decisions affecting your work,
or are you satisfied with the way things are?
%
47.8 Should have more say
52.0 Satisfied with way things are
0.2   (Don't Know)

ASK IF IN PAID WORK AT [REconAct]   n=431
Q383 [WkPrefJb]
If without having to work, you had what you would regard as a reasonable
living income, do you think you would still prefer to have a paid job
(**IF SELF-EMPLOYED:** do paid work) or wouldn't you bother?
%
79.9 Still prefer paid job (work)
19.8 Wouldn't bother
0.3   Other answer (WRITE IN)
  -   (Don't Know)

ASK IF 'employee'/DK AT [REmploye]   n=369
Q385 [PrefHour]
Thinking about the number of hours you work each
week including regular overtime, would you prefer a job
where you worked
... **READ OUT** ...
%
2.4   ... more hours per week,
28.5 fewer hours per week
69.1 or, are you happy with the number of hours you work at present?
  -   (Don't Know)

IF 'more hours' AT [PrefHour]   n=9
Q386 [MoreHour]
Is the reason why you don't work more hours because
... **READ OUT** ...
%
85.1 ... your employer can't offer you more hours,
5.8   or, your personal circumstances don't allow it?
9.2   Other answer (WRITE IN)
  -   (Don't Know)

IF 'fewer hours' AT [PrefHour]   n=105
Q388 [FewHour]
In which of these ways would you like your working hours to be
shortened ... **READ OUT** ...
%
39.3 ... shorter hours each day,
56.3 or, fewer days each week?
4.4   Other answer (WRITE IN)
  -   (Don't Know)

Q390 [EarnHour]
Would you still like to work fewer hours, if it meant earning less
money as a result?
%
23.8 Yes
73.1 No
3.1   It depends
  -   (Don't Know)

ASK IF IN PAID WORK AT [REconAct]   n=431
Q391 [WkWorkHd]
CARD
Which of these statements best describes your feelings about your job?
%
      In my job:
10.1 I only work as hard as I have to
40.6 I work hard, but not so that it interferes with the rest of my life
49.4 I make a point of doing the best I can, even if it sometimes does
      interfere with the rest of my life
  -   (Don't Know)

ASK IF 'Wholly retired' AT [REconAct]   n=113
Q392 [REmplPen]
Do you receive a pension from any past employer?
%
38.7 Yes
61.3 No
  -   (Don't Know)

IF 'married'

Q393 [SEmplPen]   n=30
Does your (husband/wife/partner) receive a pension from any past employer?
%
20.5   Yes
79.5   No
-      (Don't Know)

ASK IF 'Wholly retired' AT [REconAct]   n=113

Q394 [PrPenGet]
And do you receive a pension from any private arrangements you have made in the past, that is apart from the state pension or one arranged through an employer?
%
3.5   Yes
96.5   No
-      (Don't Know)

IF 'married'

Q395 [SPrPnGet]   n=30
And does your (husband/wife/partner) receive a pension from any private arrangements (he/she) has made in the past, that is apart from the state pension or one arranged through an employer?
%
-      Yes
100   No
-      (Don't Know)

ASK IF 'Wholly retired' AT [REconAct]   n=113

Q396 [RetAge]
%   Derived variable - over 60/65
83.9   Yes
16.1   No
-      (Don't Know)

IF 'Wholly retired' AND MALE AGED 66 OR OVER OR FEMALE AGED 61 OR OVER   n=94

Q397 [RPension]
On the whole would you say the present state pension is on the low side, reasonable, or on the high side?
IF 'ON THE LOW SIDE': Very low or a bit low?
%
43.0   Very low
34.7   A bit low
22.4   Reasonable
-      On the high side
-      (Don't Know)

25

Q398 [RPenInYr]   n=94
Do you expect your state pension in a year's time to purchase more than it does now, less, or about the same?
%
1.9   More
72.6   Less
23.2   About the same
2.3   (Don't Know)

ASK IF 'Wholly retired' AT [REconAct]   n=113

Q399 [RetirAg2]
At what age did you retire from work?
NEVER WORKED, CODE: 00
Range: 0 ... 80
Median: 60.0

HOUSING

ASK ALL

Q400 [AreaChng]   n=786
Now some questions about the area in which you live. Taking everything into account, would you say this area has got better, worse or remained about the same as a place to live during the last two years?
IF NECESSARY: By 'your area' I mean whatever you feel if your local area.

Q401 [AreaFut]   n=786
And what do you think will happen during the next two years: will this area get better, worse or remain about the same as a place to live?

	[AreaChng]	[AreaFut]
	%	%
Better	23.2	19.9
Worse	17.5	14.7
About the same	58.3	62.9
(Don't Know)	1.0	2.5

Q402 [NoisyNgb]
CARD
Please use this card to say how common or uncommon each of the following things is in your area.
Noisy neighbours or loud parties?

Q403 [Graffiti]
CARD AGAIN
(How common or uncommon is this in your area?)
Graffiti on walls and buildings?

26

n=786

Q404 [TeenOnSt]
CARD AGAIN
(How common or uncommon is this in your area?)
Teenagers hanging around on the streets?

	[NoisyNgb]	[Graffiti]	[TeenOnSt]
	%	%	%
Very common	3.8	7.1	16.8
Fairly common	5.7	12.6	20.1
Not very common	26.4	27.5	26.2
Not at all common	64.1	52.8	37.0
(Don't Know)	0.1	-	-

Q405 [Drunks]
CARD AGAIN
(How common or uncommon is this in your area?)
Drunks or tramps on the street?

Q406 [Rubbish]
CARD AGAIN
(How common or uncommon is this in your area?)
Rubbish and litter lying about?

Q407 [HmGdBad]
CARD AGAIN
(How common or uncommon is this in your area?)
Homes and gardens in bad condition?

	[Drunks]	[Rubbish]	[HmGdBad]
	%	%	%
Very common	5.1	13.8	3.2
Fairly common	8.3	18.3	8.4
Not very common	26.5	34.2	46.3
Not at all common	60.1	33.7	41.7
(Don't Know)	-	-	0.4

27

n=786

Q408 [Vandals]
CARD AGAIN
(How common or uncommon is this in your area?)
Vandalism and deliberate damage to property?

Q409 [RelTens]
CARD AGAIN
(How common or uncommon is this in your area?)
Insults or attacks to do with someone's religion?

Q410 [Burglary]
CARD AGAIN
(How common or uncommon is this in your area?)
Homes broken into?

	[Vandals]	[RelTens]	[Burglary]
	%	%	%
Very common	6.6	2.9	5.2
Fairly common	15.1	3.8	16.6
Not very common	32.7	30.1	45.9
Not at all common	45.4	63.2	31.8
(Don't Know)	0.2	0.1	0.5

Q411 [VehTheft]
CARD AGAIN
(How common or uncommon is this in your area?)
Cars broken into or stolen?

Q412 [Attacks]
CARD AGAIN
(How common or uncommon is this in your area?)
People attacked in the streets?

	[VehTheft]	[Attacks]
	%	%
Very common	8.2	1.0
Fairly common	14.5	3.9
Not very common	44.4	31.6
Not at all common	32.2	63.6
(Don't Know)	0.7	-

28

Q413 [LocTrans]   n=786
Generally speaking, would you say that compared to other areas the public transport around here is better, worse or about average?
IF 'BETTER' OR 'WORSE': Is that much (Better/Worse), or just a bit (better/worse)?

Q414 [LocEduc]
Now thinking about schools around here. Generally speaking, would you say that compared to other areas the schools around here are better, worse or about average? IF 'BETTER' OR 'WORSE': Is hat much (better/worse), or just a bit (better/worse)?

Q415 [LocJobs]
What about someone from around here applying for a job?
Generally speaking, would you say that compared to people in other areas the person's chance of being given an interview are better, worse or about average?
IF 'BETTER' OR 'WORSE': Is that much (better/worse), or just a bit (better/worse)?

	[LocTrans]	[LocEduc]	[LocJobs]
	%	%	%
Much better than average	4.6	10.3	1.5
A bit better than average	13.1	24.6	10.8
About average	54.0	58.5	70.3
A bit worse than average	8.1	1.4	7.6
Much worse than average	15.3	0.6	3.2
(Depends)	0.6	0.4	0.9
(Don't Know)	4.3	4.2	5.7

Q416 [NghBrHd]
Can I just check, how long have you lived in your present neighbourhood?
ENTER YEARS. ROUND TO NEAREST YEAR.
PROBE FOR BEST ESTIMATE.
IF LESS THAN ONE YEAR, CODE 0.
Range: 0 97
Median: 15.0

29

Q417 [HomeType]   n=786
CODE FROM OBSERVATION AND CHECK WITH RESPONDENT.
Would I be right in describing this accommodation as a ....
READ OUT ONE YOU THINK APPLIES ...
%
39.6 ... detached house or bungalow
26.6 ... semi-detached house or bungalow
29.2 ... terraced house or bungalow
3.5 ... self-contained, purpose-built flat/maisonette (inc. tenement block)
1.0 ... self-contained converted flat/maisonette
0.1 ... room(s), not self-contained
0.1 Other answer (WRITE IN)
- (Don't Know)

Q419 [NoRooms]
How many rooms does your household have for its own use?
Please exclude kitchens under 2 metres (6 feet 6 inches) wide, bathrooms, toilets and hallways.
PROMPT ON HOUSEHOLD DEFINITION IF NECESSARY
Range: 1 97
Median: 6.0

Q420 [HomeEst]
May I just check, is your home part of a housing estate?
NOTE: MAY BE PUBLIC OR PRIVATE, BUT IT IS THE RESPONDENT'S VIEW WE WANT
%
44.1 Yes, part of estate
55.9 No
- (Don't Know)

Q421 [HomeMove]
If you had a free choice, would you choose to stay in your present home, or would you choose to move out?
%
73.6 Would choose to stay
26.2 Would choose to move out
0.3 (Don't Know)

Q422 [HsePExpt]
In a year from now, do you expect house prices in your area to have gone up, to have stayed the same, or to have gone down?
IF 'GONE UP' OR 'GONE DOWN': By a lot or a little?
%
23.9 To have gone up by a lot
51.1 To have gone up by a little
18.8 To have stayed the same
1.0 To have gone down by a little
0.8 To have gone down by a lot
4.5 (Don't Know)

30

423 [HsePChng]   n=786
And compared to five years ago, would you say that house prices in your area have gone up, have stayed the same, or have gone down?
IF 'GONE UP' OR 'GONE DOWN' : By a lot or a little?
%
53.0 Have gone up by a lot
34.5 Have gone up by a little
6.7 Have stayed the same
1.0 Have gone down by a little
0.5 Have gone down by a lot
4.2 (Don't Know)

Q424 [Tenure1]
Does your household own or rent this accommodation?
PROBE IF OWNS: Outright or on a mortgage?
PROBE IF RENTS: From Whom?
%
33.6 OWNS: Own (leasehold/freehold) outright
41.4 OWNS: Buying (leasehold/freehold) on mortgage
19.4 RENTS: Housing Executive
1.1 RENTS: Housing Association
0.4 RENTS: Property company
0.3 RENTS: Employer
0.1 RENTS: Other organisation
0.1 RENTS: Relative
2.9 RENTS: Other individual
0.4 Rent free, squatting etc
- (Don't Know)

Q425 [Tenure2]
%
Derived variable
75.0 owned/being bought
19.4 Rented (Housing Executive/New Town)
1.1 Rented (Housing Association)
3.8 Rented (other)
0.4 Rent free, squatting etc
0.3 No information
- (Don't Know)

Q426 [LegalRes]   n=786
IF 'OWNS: Owns outright' AT [Tenure1]: Are the deeds for the (house/flat) in your name or are they in someone else's?
IF IN RESPONDENT'S NAME: Are they in your name only or jointly with someone else?
IF 'OWNS: Buying on a mortgage' AT [Tenure1]: Is the mortgage in your name or is it in someone else's?
IF IN RESPONDENT'S NAME: Is it in your name only or jointly with someone else?
IF 'RENTS' AT [Tenure1]: Is the rent book in your name or is in someone else's?
IF IN RESPONDENT'S NAME: Is it in your name only or jointly with someone else?
IF 'Rent free/Squatting/Outright/Other/DK/Refusal' AT [Tenure1]:
Are you legally responsible for the accommodation or is someone else?
IF LEGALLY RESPONSIBLE: Is that on your own or jointly with someone else?
%
29.5 (Deeds/Mortgage/Rent book) in respondent's name only/Yes, respondent solely responsible
43.7 Jointly with someone else
26.7 (Deeds/Mortgage/Rent book) in some else's name/No responsibility
0.1 (Don't Know)

Q427 [BuyFrmLA]   n=592
IF 'OWNS: Owns outright' or 'OWNS: Buying on mortgage' AT [Tenure1]
Did you, or the person responsible for the mortgage, buy your present home from the Housing Executive as a tenant?
%
17.9 Yes
81.7 No
- (Don't Know)
0.4 (Refusal/NA)

Q428 [CopeMorg]
IF 'OWNS: Buying on mortgage' AT [Tenure1]
How are you and your household coping with the cost of your mortgage these days? Does it make things
... READ OUT ...
%
1.5 ... very difficult,
20.0 a bit difficult,
74.1 or, not really difficult?
3.6 (Don't Know)
0.7 (Refusal/NA)

IF 'OWNS: Owns outright' OR 'OWNS: Buying on mortgage'
AT [Tenure1]

Q429 [EasySell]   n=592
CARD
If you were to put your home on the market, how easy or difficult do you think it would be to sell under present market conditions?

%
32.6 Very easy
42.8 Fairly easy
12.3 Neither easy nor difficult
7.8 Fairly difficult
2.5 Very difficult
1.6 (Don't Know)
0.4 (Refusal/NA)

IF 'RENTS' AT [Tenure1]
Q430 [RentLev1]   n=193
How would you describe the rent for this accommodation?
Would you say it was ... READ OUT ...
48.4 ... on the high side,
44.2 reasonable,
1.6 or, on the low side?
3.3 (Living rent free)
1.2 (Don't Know)
1.2 (Refusal/NA)

ASK ALL
Q431 [RentPrf1]   n=786
If you had a free choice would you choose to rent accommodation, or would you choose to buy?
%
11.3 Would choose to rent
88.2 Would choose to buy
0.5 (Don't know)

IF 'RENTS' AT [Tenure1]
Q432 [RentExp1]   n=193
And apart from what you would like, do you expect to buy a house or a flat in the next two years, or not?
IF EXPECTS TO BUY PRESENT HOUSE/FLAT, CODE 1.
%
24.9 Yes
72.1 No, do not expect to buy
1.8 (Don't Know)
1.2 (Refusal/NA)

33

Q433 [NotBuy1]   n=193
Here are some reasons people might give for not wanting to buy a home. As I read out each one, please tell me whether or not it applies to you at present?.
I could not afford a deposit

Q434 [NotBuy2]
(And does this apply or not apply to you at present)
I would not be able to get a mortgage

Q435 [NotBuy3]
(And does this apply or not apply to you at present)
It might be difficult to keep up repayments

	[NotBuy1]	[NotBuy2]	[NotBuy3]
	%	%	%
Applies	65.9	62.7	65.6
Does not apply	32.9	34.2	32.4
(Don't Know)	-	1.8	0.7
(Refusal/NA)	1.2	1.2	1.2

Q436 [NotBuy4]
(And does this apply or not apply to you at present)
I can't afford any of the properties I'd want to buy

Q437 [NotBuy5]
(And does this apply or not apply to you at present)
I do not have a secure enough job

Q438 [NotBuy8]
(And does this apply or not apply to you at present)
I might not be able to resell the property when I wanted to

	[NotBuy4]	[NotBuy5]	[NotBuy8]
	%	%	%
Applies	69.4	69.7	34.2
Does not apply	28.0	29.1	59.1
(Don't Know)	1.4	1.2	5.4
(Refusal/NA)	1.2	-	1.2

34

COMMUNITY RELATIONS

ASK ALL

Q440 [PrejRC]   n=786
Now I would like to ask some questions about religious prejudice against both Catholics and Protestants in Northern Ireland.
First thinking of Catholics - do you think there is a lot of prejudice against them in Northern Ireland nowadays, a little, or hardly any?

Q441 [PrejProt]
And now, thinking of Protestants - do you think there is a lot of prejudice against them in Northern Ireland nowadays, a little, or hardly any?

	[PrejRC]	[PrejProt]
	%	%
A lot	25.7	17.2
A little	47.5	50.6
Hardly any	22.6	27.8
(Don't Know)	3.4	3.6
(Refusal/NA)	0.8	0.8

Q442 [SRRJPrej]   n=786
How would you describe yourself ... READ OUT ...
%
0.1 ... as very prejudiced against people of other religions,
11.1 a little prejudiced,
88.0 or, not prejudiced at all?
0.2 Other (WRITE IN)
0.5 (Refusal/NA)

Q445 [RJRelAgo]
What about relations between Protestants and Catholics?
Would you say they are better than they were 5 years ago, worse, or about the same as then?
IF 'IT DEPENDS': On the whole ....
%
45.2 Better
10.0 Worse
42.8 About the same
-    Other (WRITE IN)
1.9 (Don't Know)
0.1 (Refusal/NA)

Q448 [RJRelFut]   n=786
And what about in 5 years time?
Do you think relations between Protestants and Catholics will be better than now, worse than now, or about the same as now?
IF 'IT DEPENDS': On the whole ....
%
43.3 Better than now
7.8 Worse than now
41.5 About the same
1.5 Other (WRITE IN)
5.9 (Don't Know)
0.1 (Refusal/NA)

Q451 [RelgAlwy]
Do you think that religion will always make a difference to the way people feel about each other in Northern Ireland?
%
85.8 Yes
10.3 No
2.2 Other (WRITE IN)
1.4 (Don't Know)
0.3 (Refusal/NA)

Q454 [BRelDiff]
Do you think most people in Northern Ireland would mind or not mind if s suitably qualified person of a different religion were appointed as their boss?
IF WOULD MIND: A lot or a little?

Q455 [PRelDiff]
And you personally, would you mind or not mind?
IF WOULD MIND: A lot or a little?

	[BRelDiff]	[PRelDiff]
	%	%
Would mind a lot	6.9	1.2
Would mind a little	23.7	3.2
Would not mind	63.8	94.6
(Don't Know)	5.2	0.6
(Refusal/NA)	0.4	0.4

Q456 [MarDiff]
And do you think most people in Northern Ireland would mind or not if one of their close relative were to marry someone of a different religion?
IF WOULD MIND: A lot or a little?

Q457 [MarPDiff]　　　　　　　　　　　　　　　　　　　　　　　**n=786**
And you personally, would you mind or not mine?
IF WOULD MIND: A lot or a little?

	[MarDiff]	[MarPDiff]
	%	%
Would mind a lot	17.5	7.6
Would mind a little	46.1	16.9
Would not mind	32.7	74.8
(Don't Know)	3.4	0.3
(Refusal/NA)	0.3	0.3

Q458 [MxRlgNgh]
If you had a choice, would you prefer to live in a neighbourhood with people of only own religion, or in a mixed-religion neighbourhood?
PROBE IF NECESSARY: Say if you were moving ...

%	
Own religion only	13.0
Mixed-religion neighbourhood	82.9
(Don't Know)	3.8
(Refusal/NA)	0.3

Q459 [MxRlgWrk]
And if you were working and had to change your job, would you prefer a workplace with people of only your own religion, or a mixed-religion workplace?
PROBE IF NECESSARY: Say if you did have a job ...

%	
Own religion only	3.0
Mixed-religion workplace	95.3
(Don't Know)	1.4
(Refusal/NA)	0.3

Q460 [OwnMxSch]
And if you were deciding where to send your children to school, would you prefer a school with children of only your own religion, or a mixed-religion school?
PROBE IF NECESSARY: Say if you did have school-age children ...

%	
Own religion only	30.5
Mixed-religion school	64.6
(Don't Know)	4.5
(Refusal/NA)	0.3

Q461 [JbRlgChl]　　　　　　　　　　　　　　　　　　　　　　　**n=786**
Thinking now about employment ...
On the whole, do you think the Protestants and Catholics in Northern Ireland who apply for the same job have the same chance of getting a job or are their chances of getting a job different?
IF 'IT DEPENDS' PROMPT': On the whole ...

%	
Same chance	51.0
Different chance	42.8
(Don't Know)	5.7
(Refusal/NA)	0.5

Q462 [JbRlgCh2]
Which group is more likely to get a job - Protestant or Catholics?
IF 'IT DEPENDS' PROMPT': On the whole ...

%	
Protestants	51.1
Catholics	22.8
(Don't Know)	25.0
(Refusal/NA)	1.1

ASK ALL
Q463 [ChRlgRsp]
CARD
Do you think the government and public bodies should or should not ...
... do more to teach Catholic and Protestant children greater respect for each other?

Q464 [IntegHse]
CARD
(Do you think the government and public bodies should or should not ...)
... do more to create integrated housing?

	[ChRlgRsp]	[IntegHse]
	%	%
Definitely should	72.8	54.6
Probably should	19.4	33.1
Probably should not	3.0	7.9
Definitely should not	2.3	2.0
(Don't Know)	2.0	2.1
(Refusal/NA)	0.3	0.3

Q465 [BtrComRl]
CARD AGAIN
(Do you think the government and public bodies should or should not ...)
... do more to create better community relations generally?

Q466 [IntegWrk]   n=786
CARD AGAIN
(Do you think the government and public bodies should or should not ...)

... do more to create integrated workplaces?

	[BtrComRl]	[IntegWrk]
	%	%
Definitely should	71.2	66.0
Probably should	24.6	26.4
Probably should not	1.9	4.5
Definitely should not	0.9	1.2
(Don't Know)	1.0	1.6
(Refusal/NA)	0.3	0.3

Q467 [ClosScot]
CARD
We are interested in how close people who live in Northern Ireland feel towards people living elsewhere. How close do you personally feel to

... people living in Scotland?

Q468 [ClosEng]
CARD AGAIN
How close do you personally feel to ... people living in England?

	[ClosScot]	[ClosEng]
	%	%
Very close	16.3	7.5
Fairly close	30.6	25.3
Not very close	32.6	42.5
Not at all close	19.7	24.1
(Don't Know)	0.5	0.4
(Refusal/NA)	0.2	0.2

Q469 [ClosIre]
CARD AGAIN
And how close do you personally feel to ... people living in the Republic of Ireland?

Q470 [ClosEEC]   n=786
CARD AGAIN
And how close do you personally feel to ... Europeans in General?

	[ClosIre]	[ClosEEC]
	%	%
Very close	10.9	1.9
Fairly close	43.8	15.5
Not very close	30.3	43.4
Not at all close	14.6	38.6
(Don't Know)	0.2	0.5
(Refusal/NA)	0.2	0.2

Q471 [ClosAll]
CARD
And of all these group, which one do you feel closest to?

%	
32.0	People living in Scotland
12.6	People living in England
38.9	People living in the Republic of Ireland, or
3.0	European in general?
12.3	(None of the above)
1.1	(Don't Know)
0.2	(Refusal/NA)

Q472 [ClosRel]   n=690
And do you feel closer or (answer given at ClosAll) than you do to people living in Northern Ireland who are a different religion to yourself?

%	
13.9	Yes
83.4	No
0.4	Have no religion
0.9	(Don't Know)
1.4	(Refusal/NA)

ASK ALL
Q473 [NISupPty]   n=786
Generally speaking, do you think of yourself as a supporter of any one political party?

%	
32.7	Yes
66.8	No
0.2	(Don't Know)
0.3	(Refusal/NA)

IF 'No/DK' AT [NISupPty]   **n=529**

Q475 [NIClsPty]

Do you think of yourself as a little close to one
political party than to others?

%
37.5 Yes
61.7 No
0.8 (Don't Know)

Q478 [NIPtyID1]

IF 'Yes' AT [NIClsPty]: Which one?

IF 'No/DK' AT [NIClsPty]: If there were a general election   **n=786**
tomorrow, which political party do you think you would be most
likely to support?

%   **CODE ONE ONLY**
5.2 Conservative
5.4 Labour
0.8 Liberal Democrat
7.3 Alliance (Northern Ireland)
7.8 DUP/Democratic Unionist Party
20.0 UUP/Ulster Unionist Party
1.8 Other Unionist
24.1 Sinn Fein
     SDLP
0.2 Workers Party
0.1 Campaign for Equal Citizenship
0.8 Green Party
0.7 Other Party (WRITE IN)
2.3 Other answer (WRITE IN)
15.5 None
2.9 (Don't Know)
2.9 (Refusal/NA)

IF 'Conservative', 'Labour' OR 'Liberal Democrat' AT
[NIPtyID1]   **n=135**

Q484 [NIPtyID3]

If there were a general election in which only Northern Ireland
parties were standing, which one do you think you would be most likely to
support?

%   **CODE ONE ONLY**
1.6 Other Party
3.0 Other answer
13.9 None
13.8 Alliance
1.4 DUP/Democratic Unionist Party
14.5 UUP/Ulster Unionist Party
1.9 Other Unionist party
-    Sinn Fein
6.0 SDLP
2.6 Workers Party
-    Campaign for Equal Citizenship
1.3 Green Party
5.4 (Don't Know)
34.6 (Refusal/NA)

IF NORTHERN IRELAND PARTY MENTIONED AT
[NIPtyID1] OR [NIPtyID3]   **n=615**

Q489 [NIIDStrn]

Would you call yourself very strong ... (name of Northern Ireland
party) ... fairly strong, or not very strong? Very strong (name of
Northern Ireland party)

%
7.9 Very strong (name of Northern Ireland party)
32.8 Fairly strong
50.4 Not very strong
0.1 (Don't Know)
8.8 (Refusal/NA)

ASK ALL
Q490 [NINatID]   **n=786**
CARD

%   Which of these best describes the way you think of yourself?
39.1 British
27.6 Irish
6.5 Ulster
25.1 Northern Irish
0.2 British Irish
1.0 Other (WRITE IN)
0.2 (Don't Know)
0.3 (Refusal/NA)

Q493 [UntdIrel]   n=786
At any time in the next 20 years, do you think it is likely or unlikely
that there will be a united Ireland?
PROBE: **Very likely/unlikely or quite likely/unlikely?**
%
5.1   Very likely
22.1 Quite likely
28.2 Quite unlikely
36.6 Very unlikely
4.0   (Even chance)
3.8   (Don't Know)
0.3   (Refusal/NA)

Q494 [GovIntNI]
CARD
Under direct rule from Britain, as now, how much do you generally
trust **British Governments** of any party to act in the best interests of
Northern Ireland?
PROBE IF NECESSARY

Q495 [StrIntNI]
CARD AGAIN
If there was self-rule, how much do you think you would generally
trust a **Stormont government** to act in the best interests of
Northern Ireland?
PROBE IF NECESSARY

Q496 [IrelIntNI]
CARD AGAIN
And if there was a united Ireland, how much do you think you would generally
trust an **Irish government** to act in the best interests of
Northern Ireland?
PROBE IF NECESSARY

	[GovIntNI]	[StrIntNI]	[IrelIntNI]
	%	%	%
Just about always	1.7	4.5	2.3
Most of the time	14.5	32.8	18.3
Only some of the time	41.0	30.8	37.8
Rarely	26.2	13.2	20.4
Never	14.1	14.9	16.3
(Don't Know)	2.5	3.6	4.8
(Refusal/NA)	0.1	0.1	0.1

43

COUNTRYSIDE

ASK ALL   n=786
Q497 [Outing]
Now some questions about the countryside.
Have you yourself visited the countryside or coast in the past year,
for an outing of some sort, like a drive, a walk or a picnic,
or to do something else?
IF 'Yes', PROBE: 'Was that just once, or twice or more?
%
6.7   Yes, once
83.5 Yes, twice or more
9.6   No
0.1   (Don't Know)
0.1   (Refusal/NA)

IF 'Yes, once' OR 'Yes, twice or more' AT [Outing]
Q498 [OutInCar]   n=711
When you visited the countryside in the past year, did you usually
spend ... **READ OUT** ...
%
1.5   ... all the time in the car,
9.3   almost all of the time,
39.6 only some of the time,
42.3 very little of the time,
6.6   or, did you not use a car at all?
0.5   (Depends)
 -    (Don't Know)
0.2   (Refusal/NA)

ASK ALL   n=786
Q499 [CtrySame]
Do you think the countryside generally is much the same as it was
twenty years ago, or do you think it has changed?
IF CHANGED: Has it changed a bit or a lot?
%
17.7 Much the same
26.9 Changed a bit
53.1 Changed a lot
2.1   (Don't Know)
0.1   (Refusal/NA)

IF 'Changed a bit' OR 'Changed a lot' AT [CtrySame]
Q500 [CtryBetr]   n=647
Do you think the countryside generally has changed
for the better or worse?
%
42.6 Better
40.4 Worse
13.6 (Better in some ways/worse in others)
0.6   (Don't Know)
2.7   (Refusal/NA)

44

ASK ALL

Q501 [CtryConc]   n=786

Are you personally concerned about things that may happen to the countryside, or does it not concern you particularly?

IF CONCERNED: Are you very concerned, or just a bit concerned?

%
32.4 Very concerned
36.6 A bit concerned
30.8 Does not concern me particularly
0.1   (Don't Know)
0.1   (Refusal/NA)

Q502 [CThtNew1]
      CARD
      Which, if any, of the things on this card do you think is the greatest threat to the countryside; if you think none of them is a threat, please say so.

%
25.5 Litter and fly-tipping of rubbish
17.6 New housing and urban sprawl
3.1   Superstores and out-of-town shopping centres
3.6   Building new roads and motorways
8.4   Industrial development like factories, quarries and power stations
34.1 Land and air pollution, or discharges into rivers and lakes
2.6   Changes to traditional ways of farming and of using farmland
2.3   Changes to the ordinary natural appearance of the countryside, including plants and wildlife
0.6   The number of tourists and visitors in the countryside
0.6   Other answer (**WRITE IN**)
1.3   (None of these)
0.2   (Don't Know)

IF ANSWER GIVEN AT [CThtNew1] (I.E. NOT 'None of these'/DK/Refusal)

Q504 [CThtNew2]   n=786
      CARD AGAIN
%   And which do you think is the next greatest threat?
18.5 Litter and fly-tipping of rubbish
12.5 New housing and urban sprawl
6.1   Superstores and out-of-town shopping centres
8.4   Building new roads and motorways
12.7 Industrial development like factories, quarries and power stations
24.4 Land and air pollution, or discharges into rivers and lakes
7.2   Changes to traditional ways of farming and of using farmland
5.9   Changes to the ordinary natural appearance of the countryside, including plants and wildlife
0.4   The number of tourists an visitors in the countryside
0.7   Other answer (**WRITE IN**)
2.7   (None of these)
0.1   (Don't Know)
0.2   (Refusal/NA)

ASK ALL

Q506 [FactWste]
      Suppose it is discovered that some of the factory's waste has begun leaking into a nearby river. Should the factory ... **READ OUT** ...
%
9.0   ... just be asked to do something about it,
51.5 or, should it be heavily fined for every week it continues,
39.0 or, should it be shut down unless it does something about it?
0.4   (Don't Know)
0.1   (Refusal/NA)

Q507 [PollPays]
%   Do you think the government should ... **READ OUT** ...
25.0 ... help factories meet the cost of preventing pollution,
74.3 or, should those factories that cause pollution be made to pay the bills themselves?
0.5   (Don't Know)
0.2   (Refusal/NA)

Q508 [CtryPay1]   n=786
CARD
Looking after the countryside cost a great deal of money.
From the groups on this card, please say which should have the most responsibility for footing the bill, and which the next most.
First, the most responsibility.

IF ANSWER GIVEN AT [CtryPay1] (I.E. NOT 'None of these'/DK/Refusal)
Q509 [CtryPay2]
CARD AGAIN
And which one do you feel should have the next most responsibility?

	[CtryPay1]	[CtryPay2]
	%	%
The general public through income tax and VAT	41.4	17.0
Everyone who lives in the countryside, through their council taxes	12.0	17.0
Visitors and holidaymakers through fees and charges	6.1	16.6
The farming community through their profits	7.8	13.0
Other businesses and industries in the countryside, through their profits	28.8	29.1
(None of these)	2.3	4.5
(Don't Know)	1.5	1.3
(Refusal/NA)	0.1	1.6

ASK ALL
Q510 [ResPres]
Can I just check, would you describe the place where you live as ... READ OUT ...
%
6.2   ... a big city,
21.1   the suburbs or outskirts of a big city,
37.1   a small city or town,
15.0   a country village
20.1   or, a farm or home in the countryside?
0.3   (Other answer (WRITE IN)
0.2   (Don't Know)
0.1   (Refusal/NA)

47

Q513 [MotorVPT]   n=786
If over the next few years the government has to choose, to which should it give greater priority ... READ OUT ...
%
22.6   ... to needs of motorists,
69.3   or, the needs of public transport users?
1.3   (Neither)
5.6   (Both)
1.1   (Don't Know)
0.1   (Refusal/NA)

POLITICAL TRUST

ASK ALL
Q515 [GovtWork]
CARD
Which of these statements best describes your opinion on the present system of governing in UK?
%
2.0   Works extremely well and could not be improved
25.0   Could be improved in small ways but mainly works well
46.4   Could be improved quite a lot
24.6   Needs a great deal of improvement
1.8   (Don't Know)
0.1   (Refusal/NA)

Q516 [Lords]
Do you think that the House of Lords should remain as it is or is some changes needed?
%
25.0   Remain as it is
61.1   Change needed
13.6   (Don't Know)
0.4   (Refusal/NA)

IF 'Change needed' AT[Lords]
Q517 [LordsHow]   n=590
Do you think the House of Lords should be ... READ OUT ...
%
17.1   ... replace by a different body
26.0   abolished and replaced by nothing
36.2   or, should there be some other kind of change?
2.1   (Don't Know)
18.6   (Refusal/NA)

48

ASK ALL

Q518 [Monarchy]   n=786

How about the monarchy or the Royal family in the UK.
How important or unimportant do you think it is for the UK to continue to have a monarchy ... **READ OUT** ...

%
19.1 ... very important
31.2 quite important
20.7 not very important
12.4 not at all important
14.2 or, do you think the monarchy should be abolished?
1.9 (Don't Know)
0.3 (Refusal/NA)

Q519 [Coalitn]

Which do you think is generally better in the UK
... **READ OUT** ...

%
35.7 ... to have a government formed by one political party
58.0 or, for two or more parties to get together to form a government?
6.0 (Don't Know)
0.3 (Refusal/NA)

Q520 [VoteSyst]

Some people say that we should change the voting system to allow smaller political parties to get a fairer share of MPs. Others say we should keep the voting system as it is, to produce more effective government. Which view comes closest to your own
... **READ OUT** ...

IF ASKED, REFERS TO 'PROPORTIONAL REPRESENTATION'

%
43.2 ... that we should change the voting system
51.1 or, keep it as it is?
5.4 (Don't Know)
0.3 (Refusal/NA)

49

Q521 [ScotPar1]   n=786
CARD
An issue in Scotland is the question of an elected Assembly - a special parliament for Scotland dealing with Scottish affairs. Which of these statements comes closest to your view?

%
7.1 Scotland should become independent, separate from the UK and the European Union
14.2 Scotland should become independent, separate from the UK, but part of the European Union
46.3 Scotland should remain part of the UK but with its own elected Assembly that has some taxation and spending powers
22.0 There should be no change from the present system
1.1 (Other answer (**WRITE IN**))
9.0 (Don't Know)
0.3 (Refusal/NA)

Q523 [NIreland]

Do you think the long-term policy for Northern Ireland should be for it ... **READ OUT** ...

62.1 ... to remain part of the United Kingdom
23.9 or, to unify with the rest of Ireland?
1.7 Independent State
0.2 Irish should decide
6.2 (Other answer (**WRITE IN**))
5.6 (Don't Know)
0.3 (Refusal/NA)

Q525 [DecFutNI]

And who do you think should have the right to decide what the long-term future of Northern Ireland should be?

%
63.4 ... the people in Northern Ireland on their own
19.9 or, the people of Ireland, both north and south
10.3 or, the people both in Northern Ireland and in Britain?
2.7 or, the people of Great Britain, Northern Ireland and Eire
1.0 (Other answer (**WRITE IN**))
2.4 (Don't Know)
0.2 (Refusal/NA)

50

Q527 [TroopOut]
n=786
Some people say that government policy towards Northern Ireland should include a complete withdrawal of British troops. Would you personally support or oppose such a policy?
IF 'SUPPORT' OR 'OPPOSE', PROBE: Strongly or a little?

%	
14.9	Support strongly
16.4	Support a little
23.7	Oppose a little
36.7	Oppose strongly
3.3	Other answer (WRITE IN)
0.1	(Troops should be withdrawn in the long-term but not immediately)
0.1	(It should be up to the Irish to decide)
4.2	(Don't Know)
0.8	(Refusal/NA)

Q529 [ECPolicy]
CARD
Do you think the UK's long-term policy should be
... **READ OUT** ...

%	
11.2	... to leave the European Union,
28.6	to stay in the EU and try to reduce the EU's powers,
23.5	to leave things as they are,
13.8	to stay in the EU and try to increase the EU's powers,
11.9	or, to work for the formation of a single European government?
10.5	(Don't Know)
0.6	(Refusal/NA)

Q530 [EcuView]
CARD
And here are three statements about the future of the pound in the European Union. Which one comes closest to your view?

%	
23.9	Replace the pound by a single currency
14.1	Use both the pound and a new European currency in the UK
57.2	Keep the pound as the only currency for the UK
4.6	(Don't Know)
0.2	(Refusal/NA)

Q531 [ObeyLaw]
In general, would you say that people should obey the law without exception, or are there exceptional occasions on which people should follow their consciences even if it means breaking the law?

%	
50.7	Obey law without exception
46.8	Follow conscience on occasions
2.3	(Don't Know)
0.2	(Refusal/NA)

Q532 [DefnScrt]
n=786
Do you think that the government should have the right to keep its defence plans secret or do you think the public should normally have the right to know what they are?

Q533 [EconScrt]
And what about its economic plans? Should the government have the right to keep these secret or should the public normally have the right to know what they are?

Q534 [LawsScrt]
And what about its plans for new laws it is thinking of introducing? Should the government have the right to keep these secret or should the public normally have the right to know what they are?

	[DefnScrt] %	[EconScrt] %	[LawsScrt] %
Government should have the right to keep them secret	46.2	9.2	5.6
Public should normally have the right to know what they are	50.0	88.1	92.3
(Don't Know)	3.3	2.5	1.9
(Refusal/NA)	0.5	0.2	0.2

Q535 [GovNoSay]
CARD
Please choose a phrase from this card to say how much you agree or disagree with the following statements.
People like me have no say in what the government does

Q536 [LoseTch]
CARD AGAIN
(Please choose a phrase from this card to say how much you agree or disagree with this statement)
Generally speaking those we elect as MPs lose touch with people pretty quickly

Q537 [VoteIntr]   n=786
CARD AGAIN
(Please choose a phrase from this card to say how much you agree or disagree with this statement)
Parties are only interested in people's votes, not in their opinions

	[GovNoSay]	[LoseTch]	[VoteIntr]
	%	%	%
Agree strongly	35.2	31.0	33.6
Agree	37.9	49.1	43.3
Neither agree nor disagree	10.2	9.8	13.4
Disagree	14.7	8.2	8.0
Disagree strongly	1.7	0.7	0.5
(Don't Know)	0.1	0.9	1.0
(Refusal/NA)	0.2	0.2	0.2

Q538 [VoteOnly]
CARD AGAIN
(Please choose a phrase from this card to say how much you agree or disagree with this statement)
Voting is the only way people like me can have any say about how the government runs things

Q539 [GovComp]
CARD AGAIN
(Please choose a phrase from this card to say how much you agree or disagree with this statement)
Sometimes politics and government seem so complicated that a person like me cannot really understand what is going on.

Q540 [PtyNMat]
CARD AGAIN
(Please choose a phrase from this card to say how much you agree or disagree with this statement)
It doesn't really matter which party is in power, in the end things go on much the same.

	[VoteOnly]	[GovComp]	[PtyNMat]
	%	%	%
Agree strongly	22.1	26.3	22.8
Agree	50.9	50.6	52.8
Neither agree or disagree	11.5	7.0	5.6
disagree	11.9	14.0	16.6
Disagree strongly	2.4	1.2	1.3
(Don't Know)	1.0	0.7	0.6
(Refusal/NA)	0.2	0.2	0.2

53

Q541 [InfPolit]   n=786
CARD AGAIN
(Please choose a phrase from this card to say how much you agree or disagree with this statement)
I think I am better informed than most people about politics and government

Q542 [MPsCare]
CARD AGAIN
(Please choose a phrase from this card to say how much you agree or disagree with this statement)
MPs don't care much about what people like me think.

	[InfPolit]	[MPsCare]
	%	%
Agree strongly	2.8	21.1
Agree	16.4	47.8
Neither agree nor disagree	25.7	15.1
Disagree	45.8	14.0
Disagree strongly	8.6	0.7
(Don't Know)	0.2	1.1
(Refusal/NA)	0.4	0.2

Q543 [GovTrust]
CARD
How much do you trust UK governments of any party to place the needs of the nation above the interests of their own political party? Please choose from the card.

Q544 [ClrTrust]
CARD AGAIN
And how much to you trust local councillors of any party to place the needs of their area above the interests of their own political party

Q545 [PolTrust]
CARD AGAIN
How much do you trust police not the bend the rules in trying to get a conviction?

	[GovTrust]	[ClrTrust]	[PolTrust]
	%	%	%
Just about always	2.5	3.3	8.6
Most of the time	16.5	26.0	35.0
Only some of the time	50.9	51.0	37.9
Almost never	27.9	17.4	14.5
(Don't Know)	2.0	2.0	3.5
(Refusal/NA)	0.2	0.2	0.5

54

n=786

Q546 [CSTrust]
CARD AGAIN
And how much do you trust top civil servants to stand firm against a minister who wants to provide false information to parliament?

Q547 [MPsTrust]
CARD AGAIN
And how much do you trust politicians of any party in the UK to tell the truth when they are in a tight corner?

Q548 [JugTrust]
CARD AGAIN
And how much do you trust high court judges to stand up to a government which wishes them to reach a particular verdict?

	[CSTrust]	[MPsTrust]	[JugTrust]
	%	%	%
Just about always	5.2	1.2	10.2
Most of the time	23.9	7.2	34.4
Only some of the time	41.9	35.2	34.3
Almost never	21.6	53.6	14.6
(Don't know)	6.9	2.3	6.1
(Refusal/NA)	0.4	0.4	0.2

Q549
CARD
Now thinking of MPs, which of the following qualities shown on this card would you say are important for an MP to have? You may choose more than one, none, or suggest others.
Multicoded (Maximum of 9 codes)

%		
51.7	To be well educated	[MPEd]
35.8	To know what being poor means	[MPPoor]
19.4	To have business experience	[MPbus]
8.1	To have trade union experience	[MPUnion]
61.8	To have been brought up in the area he or she represents	[MPLocal]
39.6	To be loyal to the party he or she represents	[MPLoyal]
45.2	To be independent minded	[MPInd]
2.3	(To be honest/trustworthy/open)	[MPOth8]
0.6	(To be caring/compassionate)	[MPOth9]
0.4	(To be loyal to their constituents)	[MPOth10]
0.1	(To be upright/moral)	[MPOth11]
-	(To be courageous)	[MPOth12]
0.2	(To be prepared to listen/keep in touch)	[MPOth13]
1.6	None of these qualities	[MPNone]
2.3	Other important qualities (PLEASE SPECIFY)	[MPOth2]
1.5	(Don't Know)	
0.2	(Refusal/NA)	

55

CLASSIFICATION

ASK ALL
Q104 [NumAdult]   **n=786**
INTERVIEWER: YOU ARE GOING TO ASK ABOUT ALL THE ADULTS AGED 18 OR OVER IN THE HOUSEHOLD. STARTING WITH THE HOH, LIST ALL ADULTS IN DESCENDING ORDER OF AGE.
Including yourself, how many adults are there in your household, that is, people aged 18 and over whose main residence this is and who are catered for by the same person as yourself or share living accommodation with you?
ENTER NUMBER OF PERSONS AGED 18 AND OVER
Range: 1 ... 10
MEDIAN: 2.0

Q166 [NumChild]
And how many people are there in your household aged **under 18** (INCLUDING CHILDREN)
Range: 0 ... 10
MEDIAN: 0.0

dv   [HouseHld]
Range: 1 ... 20
MEDIAN: 3.0

dv   [RMarStat]
%   RESPONDENT'S MARITAL STATUS
60.6   Married
1.8   Cohabiting
22.4   Single, no children
2.1   Single parent
7.5   Widowed
2.5   Divorced after marrying
2.9   Separated after marrying
-      (Don't Know)
0.1   (Refusal/NA)

dv   [Rsex]
%   RESPONDENT'S SEX
45.0   Male
55.0   Female
-      (Don' Know)

56

dv [Rage]
RESPONDENT'S AGE   n=786
Range: 18 ... 97
MEDIAN: 42.0

Q553 [Religion]
Do you regard yourself as belonging to any particular religion?
IF YES: Which?
% CODE ONE ONLY - DO NOT PROMPT
13.6 No religion
3.2 Christian - no denomination
36.8 Roman Catholic
17.5 Church of Ireland/Anglican
1.1 Baptist
3.2 Methodist
19.0 Presbyterian/Church of Scotland
0.2 Other Christian
0.2 Hindu
- Jewish
- Islam/Muslim
- Sikh
- Buddhist
- Other non-Christian
1.2 Free Presbyterian
0.5 Brethren
0.5 United Reform Church (URC)/Congregational
2.4 Other Protestant
0.2 (Don't Know)
0.4 (Refusal/NA)

IF NOT 'Refusal/NA' AT [Religion]   n=786
Q559 [FamRelig]
In what religion, if any, were you brought up?
PROBE IF NECESSARY: What was your family's religion?
CODE ONE ONLY - DO NOT PROMPT
%
1.6 No religion
1.5 Christian - no denomination
43.4 Roman Catholic
20.5 church of Ireland/Anglican
1.3 Baptist
4.3 Methodist
23.0 Presbyterian/Church of Scotland
0.2 Other Christian
- Hindu
- Jewish
- Islam/Muslim
- Sikh
- Buddhist
- Other non-Christian
1.0 Free Presbyterian
1.0 Brethren
0.5 United Reform Church (URC)/Contregational
0.6 Other Protestant
0.3 (Don't Know)
0.7 (Refusal/NA)

IF RELIGION GIVEN AT EITHER [Religion] OR AT
[FamRelig] (I.E. NOT 'No religion' OR 'Refusal/NA' AT BOTH)
Q564 [ChAttend]   n=774
Apart from such special occasions as weddings, funerals and baptisms, how often nowadays do you attend services or meetings connected with your religion?
PROBE AS NECESSARY
%
49.2 Once a week or more
7.0 Less often but at least once in two weeks
9.8 Less often but at least once a month
9.5 Less often but at least twice a year
2.9 Less often but at least once a year
1.4 Less often
19.0 Never or practically never
0.6 Varies too much to say
- (Don't Know)
0.6 (Refusal/NA)

ASK ALL   n=786
Q565 [RaceOri2]
CARD
% To which of these groups do you consider you belong?
0.1 BLACK: of African origin
- BLACK: of Caribbean origin
-- BLACK: of other origin
- ASIAN: of Indian origin
- ASIAN: of Pakistani origin
- ASIAN: of Bangladeshi origin
- ASIAN: of Chinese origin
- ASIAN: of other origin (WRITE IN)
99.1 WHITE: of any European Origin
0.1 WHITE: of other origin (WRITE IN)
- MIXED ORIGIN (WRITE IN)
0.2 OTHER (WRITE IN)
0.1 (Don't Know)
0.2 (Refusal/NA)

Q591 [NumChild]
Derived variable - number of respondent's children aged 5 or over in household
Range: 0 ... 97
Median: 0.0

Q592 [SifMxSch]
Did you ever attend a mixed ro integrated school, that is a school with fairly large numbers of both Catholic and Protestant children?
IF YES: In Northern Ireland or Somewhere else?
%
20.4 Yes, in Northern Ireland
2.2 Yes, somewhere else
77.0 No, did not
0.4 (Refusal/NA)

IF 'None' AT [NumChild]   n=443
Q593 [OthChld3]
Have you ever been responsible for bringing up any children of school age, including stepchildren?
%
38.2 Yes
61.1 No
0.2 (Don't Know)
0.5 (Refusal/NA)

IF RESPONDENT BEEN RESPONSIBLE FOR SCHOOL AGE CHILDREN AT [NumCh] OR [OthChld3]
Q594 [ChdMxSch]   n=515
And (has/have) (any of) your child (ren) ever attended a mixer or integrated school, with fairly large numbers of both Catholics and Protestants attending?
IF YES: In Northern Ireland
%
24.9 Yes, in Northern Ireland
2.2 Yes, somewhere else
71.5 No, did not
0.1 (Don't Know)
1.3 (Refusal/NA)

ASK ALL   n=786
Q597 [TEA]
How old were you when you completed your continuous full-time education?
%
32.1 15 or under
30.0 16
10.6 17
9.2 18
14.6 19 or over
1.5 Still at school
1.6 Still at college or university
- Other answer (WRITE IN)
0.2 (Don't Know)
0.3 (Refusal/NA)

Q598 [SchQual]
CARD
Have you passed any of the examinations on this card?
%
49.0 Yes
50.6 No
0.1 (Don't Know)
0.2 (Refusal/NA)

IF 'Yes' AT [SchQual]
Q599
CARD AGAIN
Which ones?
PROBE: Any others?

n=388

	Has quals	Not answered
	%	%
CSE Grades 2 - 5		
GCSE Grades D - G	21.2	0.8
CSE Grade 1		
GCSE 'O' Level		
GCSE Grades A - C		
School certificate or matriculation	85.8	0.8
Scottish (SCE) Ordinary		
Scottish School-leaving certificate lower grade		
SUPE Ordinary		
Northern Ireland Junior Certificate		
GCE 'A' level/ 'S' level	41.7	0.8
Higher School certificate		
Scottish SCE/SLC/SUPE at Higher grade		
Northern Ireland Senior Certificate		
Overseas school leaving exam or certificate	0.9	0.8

ASK ALL
Q604 [PSchQual]
CARD
And have you passed any of the exams or got any of the qualifications on this card?

n=786

%
45.9 Yes
53.8 No
0.1 Don't Know
0.2 (Refusal/NA)

IF 'Yes' AT 'PSchQual'
Q605 [PSchQFW]
CARD AGAIN Which ones?
PROBE: Any others?

n=363

%		
13.0	Recognised trade apprenticeships completed	[EdQual5]
30.0	RSA/other clerical, commercial qualification	[EdQual6]
17.2	City & Guilds Certificate - Craft/Intermediate/Ordinary/Part 1	[EdQual7]
9.2	City & Guilds Certificate - Advanced/final/Part II or Part III	[EdQual8]
3.4	City & Guilds Certificate - Full technological	[EdQual9]
10.5	BEC/TEC General/Ordinary National Certificate (ONC) or Diploma (OND)	[EdQual10]
5.6	BEC/TEC Higher/Higher National Certificate (HNC) or Diploma (HND)	[EdQual11]
3.0	NVQ/SVQ Level 1/GNVQ Foundation level	[EdQual17]
3.2	NVQ/SVQ Level 2/GNVQ Intermediate level	[EdQual18]
2.0	NVQ/SVQ Level 3/GNVQ Advanced level	[EdQual19]
0.1	NVQ/SVQ Level 4	[EdQual20]
-	NVQ/SVQ Level 5	[EdQual21]
9.2	Teacher training qualification	[EdQual12]
7.2	Nursing qualification	[EdQual13]
8.5	Other technical or business qualification/certificate	[EdQual14]
19.3	University or CNAA degree or diploma	[EdQual15]
6.4	Other recognised academic or vocational qualification (WRITE IN)	[EdQual16]
0.9	(Don't Know)	

OCCUPATIONAL DETAILS OF SPOUSE ASKED HERE - MIRRORS 'REconAct] TO [RPartFul]

ASK IF MARRIED OR COHABITING n=491
Q684 [ReligSam]
Is your (husband/wife/partner) the same religion as you?
PROBE IF NECESSARY
%
90.0 Yes, same religion
6.3 No, not same religion
2.7 No religion at all
0.3 (Don't Know)
0.7 (Refusal/NA)

ASK ALL
Q686 [CarOwn]   n=786
Do you, or does anyone else in your household, own or have the regular use of a car or van?
%
79.7 Yes
19.9 No
0.1 (Don't Know)
0.2 (Refusal/NA)

Q687 [AnyBN2]
CARD
Do you (or your husband/wife/partner) receive any of the state benefits on this card at present?
%
67.3 Yes
32.3 No
0.1 (Don't Know)
0.4 (Refusal/NA)

IF 'Yes' AT [AnyBn2]
Q688 CARD AGAIN Which ones?   n=532
PROBE: Any others?

		Receive %	NA %
Retirement pension (National Insurance)	[BenefOap]	16.6	0.7
War Pension (War Disablement Pension or War Widows Pension)	[BenefWar]	0.6	0.7
Widow's Benefits (Widow's Pension and Widowed Mother's Allowance)	[BenefWid]	4.5	0.7
Unemployment Benefit / Income Support for the Unemployed (Jobseekers Allowance)	[BenefUB]	8.2	0.7
Income Support (Other than for employment)	[BenefIS]	16.6	0.7
Child Benefit (formerly Family Allowance)	[BenefCB]	53.8	0.7
One Parent Benefit	[BenefOB]	4.3	0.7
Family Credit	[BenefFC]	3.2	0.7
Housing Benefit (Rent Rebate)	[BenefHB]	11.7	0.7
Rate Rebate	[BenefRR]	2.3	0.7
Incapacity Benefit/Sickness Benefit/Invalidity Benefit	[BenefInc]	11.8	0.7
Disability Living Allowance	[BenefDLA]	9.0	0.7
Attendance Allowance	[BenefAtA]	3.8	0.7
Severe Disablement Allowance	[BenefSev]	0.7	0.7
Invalid Care Allowance	[BenefICA]	1.9	0.7
Other state benefit (WRITE IN)	[BenefOth]	0.8	0.7

ASK ALL
Q691 [MainInc]   n=786
CARD
Which of these is the main source of income for you (and your husband/wife/partner) at present?
%
60.9 Earnings from employment (own or spouse/partner's)
2.5 Occupational pension (s) - from previous employer (s)
15.1 State retirement or widow's pension (s)
1.4 Unemployment benefit
9.6 Income Support
0.3 Family Credit
5.1 Invalidity, sickness or disabled pension or benefit (s)
- Other state benefit (WRITE IN)
0.8 Interest from savings or investments
1.1 Student grant
2.1 Dependent on parents/other relatives
0.7 Other main source
0.1 (Don't Know)
0.4 (Refusal/NA)

Q694 [HHIncome]   n=786
CARD
Which of the letters on this card represent the total income of your household from all sources before tax?
Please just tell me the letter.
NOTE: INCLUDES INCOME FROM BENEFITS, SAVINGS, ETC.

ASK ALL RESPONDENTS WHO ARE WORKING (IF 'In paid Work' AT [REconActI])
Q695 [Rearn]   n=431
CARD AGAIN
Which of the letters on this card represents your own gross or total earnings, before deduction of income tax and national insurance?

	[HHIncome]	[REarn]
	%	%
Q	5.8	8.4
T	10.9	8.6
O	8.3	10.6
K	4.8	12.8
L	7.0	14.1
B	6.9	10.3
Z	5.4	6.7
M	5.0	4.9
F	6.5	5.4
J	5.4	3.6
D	3.1	1.0
H	2.4	0.9
C	4.0	2.0
G	1.9	0.8
P	1.5	1.3
N	5.3	1.3
(Don't Know)	11.4	2.0
(Refusal/NA)	4.5	5.3

ASK ALL
Q696 [OwnShare]   n=786
Do you (or your husband/wife/partner) own any share quoted on the Stock Exchange, including unit trusts?
14.4 Yes
84.8 No
0.3 (Don't Know)
0.5 (Refusal/NA)

65

Q697 [EvrLivGB]   n=786
% Have you ever lived in mainland Britain for more than a year?
16.1 Yes
83.5 No
0.1 (Don't Know)
0.3 (Refusal/NA)

Q698 [EvrLivEr]
% Have you ever lived in the Republic of Ireland for more than a year?
4.5 Yes
95.1 No
0.1 (Don't Know)
0.3 (Refusal/NA)

Q699 [PastVot]
Thinking back to the last general election in 1992 - do you remember which party you voted for then, or perhaps you didn't note in that election?
IF NECESSARY, SAY: The one where John Major won against Neil Kinnock
DO NOT PROMPT

%	Yes, voted:	
2.8		Conservative
1.1		Labour
-		Liberal Democrat
6.7		Alliance (Northern Ireland)
7.7		DUP/Democratic Unionist Party)
16.3		UUP/Ulster Unionist Party)
0.6		Other Unionist party
1.7		Sinn Fein
18.3		SDLP
0.1		Workers Party
-		Campaign for Equal Citizenship
-		Green Party
4.1		Other (WRITE IN)
35.9		Refused to say
2.5		Did not vote
2.0		(Can't remember/Don't know)
		(Don't Know)

66

IF 'Did not vote' AT [PastVot]

Q701 [WhNtVot]   n=318

CARD

Which of the reasons on this card comes closest to explaining why you did not manage to vote in the last general election in 1992?

%

16.7 I was not eligible or not registered to vote

7.4   I wanted to vote but was not able to get to the polling station

2.7   I didn't understand enough about politics

32.2 I was not interested enough in the election

0.9   I wanted to vote but forgot to

1.8   I never vote for reasons of conscience

3.9   I couldn't decide who to vote for

20.6 There was no one I wanted to vote for

7.7   Other reason (WRITE IN)

0.7   (Don't Know)

5.4   (Refusal/NA)

ASK ALL

Q703 [UniNatID]   n=786

Generally speaking, do you think of yourself as a unionist, a nationalist or neither?

%

36.6 Unionist

19.0 Nationalist

42.8 Neither

0.3   (Don't Know)

1.3   (Refusal/NA)

IF 'Unionist' OR 'Nationalist' AT [UniNatID]

Q704 [UniNatSt]

Would you call yourself a very strong (Unionist/Nationalist), fairly strong, or not very strong?

%

11.9 Very strong

46.0 Fairly strong

39.4 Not very strong

2.7   (Don't Know)

NI

P.1525

Spring 1996

NORTHERN IRELAND SOCIAL ATTITUDES 1996

SELF-COMPLETION QUESTIONNAIRE

OFFICE USE ONLY

6-8	Cluster number
9-13	Spare
14-15	5 0	Card no.
16-18	Spare
27-31	Batch no.
32-34	Spare

INTERVIEWER TO ENTER

1-5	8	Serial number
19-22	0	Sampling point
23-26		Interviewer number

To the selected respondent:

Thank you very much for agreeing to take part in this important study - the seventh in this annual series. The study consists of this self-completion questionnaire, and the interview you have already completed. The results of the survey are published in a book each autumn; some of the questions are also being asked in twenty-four other countries, as part of an international survey.

Completing the questionnaire:

The questions inside cover a wide range of subjects, but each one can be answered simply by placing a tick ( ) or a number in one or more of the boxes. No special knowledge is required: we are confident that everyone will be able to take part, not just those with strong views or particular viewpoints. The questionnaire should not take very long to complete, and we hope you will find it interesting and enjoyable. Only you should fill it in, and not anyone else at your address. The answers you give will be treated as confidential and anonymous.

Returning the questionnaire:

Your interviewer will arrange with you the most convenient way of returning the questionnaire. If the interviewer has arranged to call back for it, please fill it in and keep it safely until then. If not, please complete it and post it back in the pre-paid, addressed envelope, AS SOON AS YOU POSSIBLY CAN.

THANK YOU AGAIN FOR YOUR HELP.

Social and Community Planning Research is an independent social research institute registered as a charitable trust. Its projects are funded by government departments, local authorities, universities and foundations to provide information on social issues in the UK. This survey series is funded mainly by one of the Sainsbury Family Charitable Trusts, with contributions also from other grant-giving bodies and government departments. Please contact us if you would like further information.

N10

2.01 In general, would you say that people should obey the law without exception, or are there exceptional occasions on which people should follow their consciences even if it means breaking the law?

n=820

PLEASE TICK ONE BOX ONLY
OR

	%
Obey the law without exception	46.2
Follow conscience on occasions	48.6
Can't choose	3.5
NA	1.8

2.02 There are many ways people or organisations can protest against a government action they strongly oppose. Please show which you think should be allowed and which should not be allowed by ticking a box on each line.

PLEASE TICK ONE BOX ON EACH LINE

Should it be allowed?

	Definitely	Probably	Probably not	Definitely not	Can't choose	NA
a. Organising public meetings to protest against the government	% 44.9	36.0	5.0	7.0	4.9	2.3
b. Organising protest marches and demonstrations	% 23.8	32.5	15.1	18.6	4.1	5.8
c. Organising a nationwide strike of all workers against the government	% 11.8	14.9	24.4	37.9	5.7	5.2

2.03 Would you or would you not do any of the following to protest against a government action you strongly opposed?

PLEASE TICK ONE BOX ON EACH LINE

	Definitely would	Probably would	Probably would not	Definitely would not	Can't choose	NA
a. Attend a public meeting organised to protest against the government	% 24.4	37.7	18.5	13.9	4.2	1.3
b. Go on a protest march or demonstration	% 12.1	24.7	27.9	28.3	2.1	5.0

2.04 And in the past five years how many times have you done each of the following to protest against a government action you strongly oppose?

PLEASE TICK ONE BOX ON EACH LINE

	Never	Once	More than once	NA
a. Attended a public meeting organised to protest against the government	% 82.7	10.0	6.3	1.1
b. Gone on a protest march or demonstration	% 80.4	9.7	5.8	4.1

N11

2.05 There are some people whose views are considered extreme by the majority. Consider people who want to overthrow the government by revolution. Do you think such people should be allowed to ...

n=820

PLEASE TICK ONE BOX ON EACH LINE

	Definitely	Probably	Probably not	Definitely not	Can't choose	NA
a. ... hold public meetings to express their views?	% 12.8	30.9	19.5	30.6	5.1	1.2
b. ... publish books expressing their views?	% 17.1	35.8	15.7	21.6	5.7	4.0

2.06 All systems of justice make mistakes, but which do you think is worse ...

%

PLEASE TICK ONE BOX ONLY
OR

	%
... to convict an innocent person,	66.2
to let a guilty person go free?	19.7
Can't choose	13.5
NA	0.6

2.07 The government has a lot of different pieces of information about people which computers can bring together very quickly. Is this ...

%

PLEASE TICK ONE BOX ONLY

	%
... a very serious threat to individual privacy,	28.1
a fairly serious threat,	31.6
not a serious threat,	28.8
or, not a threat at all to individual privacy?	6.1
Can't choose	5.3
NA	0.1

2.08 What is your opinion of the following statement:
"It is the responsibility of the government to reduce the differences in income between people with high incomes and those with low incomes."

%

PLEASE TICK ONE BOX ONLY

	%
Agree strongly	29.2
Agree	34.4
Neither agree nor disagree	16.9
Disagree	11.7
Disagree strongly	3.6
Can't choose	3.7
NA	0.5

N12

n=820

2.09 Here are some things the government might do for the economy. Please show which actions you are in favour of and which you are against.

PLEASE TICK ONE BOX ON EACH LINE

	Strongly in favour of	In favour of	Neither in favour of nor against	Against	Strongly against	Can't Choose	NA
a. Control of wages by law	% 14.9	24.4	27.2	25.5	4.8	0.2	3.0
b. Control of prices by law	% 21.5	37.2	18.6	17.2	2.3	0.2	3.1
c. Cuts in government spending	% 17.7	28.9	20.0	22.6	5.9	0.1	4.7
d. Government financing of projects to create new jobs	% 44.0	45.6	6.7	1.1	0.4	-	2.3
e. Less government regulation of business	% 8.7	26.8	42.3	15.0	2.6	0.2	4.4
f. Support for industry to develop new products and technology	% 30.2	53.4	10.8	1.4	0.7	0.1	3.3
g. Support for declining industries to protect jobs	% 23.2	40.4	21.2	9.9	1.1	0.1	4.0
h. Reducing the working week to create more jobs	% 10.7	26.9	30.3	26.0	3.8	0.2	2.1

2.10 Listed below are various areas of government spending. Please show whether you would like to see more or less government spending in each area. Remember that if you say "much more", it might require a tax increase to pay for it.

PLEASE TICK ONE BOX ON EACH LINE

	Spend much more	Spend more	Spend the same as now	Spend less	Spend much less	Can't choose	NA
a. The environment	% 9.8	37.0	38.8	4.6	0.9	2.8	6.0
b. Health	% 51.9	41.0	5.5	0.3	0.1	0.2	0.9
c. The police and law enforcement	% 8.0	27.9	44.1	10.1	3.4	1.5	5.1
d. Education	% 39.6	43.0	13.5	0.8	0.7	0.4	2.0
e. The military and defence	% 3.0	10.3	40.2	23.3	15.5	2.7	5.1
f. Old age pensions	% 41.3	43.4	11.9	0.7	-	0.2	2.4
g. Unemployment benefits	% 15.9	26.8	34.3	14.2	4.7	1.4	2.7
h. Culture and the arts	% 1.6	7.2	29.9	32.1	22.5	3.3	3.5

N13

n=820

2.11a Do you think that trade unions in this country have too much power or too little power?

PLEASE TICK *ONE* BOX ONLY

%

Far too much power	4.5
Too much power	10.8
About the right amount of power	45.8
Too little power	22.9
Far too little power	2.8
Can't choose	13.3

b How about business and industry? Do they have too much power or too little power?

PLEASE TICK *ONE* BOX ONLY

%

Far too much power	6.1
Too much power	28.0
About the right amount of power	41.5
Too little power	9.3
Far too little power	0.1
Can't choose	14.8
NA	0.2

c And what about the government, does it have too much power or too little power?

PLEASE TICK *ONE* BOX ONLY

%

Far too much power	19.5
Too much power	31.3
About the right amount of power	39.4
Too little power	3.7
Far too little power	0.4
Can't choose	5.7

N14

2.12 On the whole, do you think it should or should not be the government's responsibility to ...

PLEASE TICK ONE BOX ON EACH LINE

m=620

	Definitely should be	Probably should be	Probably should not be	Definitely should not be	Can't choose	NA
a. ... provide a job for everyone who wants one	% 37.0	40.2	12.5	5.4	3.2	1.7
b. ... keep prices under control	% 48.0	40.0	6.2	1.2	1.3	3.3
c. ... provide health care for the sick	% 82.3	15.3	0.8	0.1	0.7	0.9
d. ... provide a decent standard of living for the old	% 80.5	16.0	0.7	0.4	1.4	1.0
e. ... provide industry with the help it needs to grow	% 39.0	46.2	7.3	1.1	3.0	3.5
f. ... provide a decent standard of living for the unemployed	% 33.8	46.5	9.2	3.8	3.5	3.1
g. ... reduce income differences between the rich and the poor	% 39.3	33.2	12.5	8.3	3.9	2.8
h. ... give financial help to university students from low-income families	% 49.1	40.0	5.1	1.6	2.0	2.2
i. ... provide decent housing for those who can't afford it	% 45.6	41.5	5.9	1.7	2.7	2.6
j. ... impose strict laws to make industry do less damage to the environment	% 62.2	28.3	3.5	1.3	2.5	2.3

Now some questions about politics.

2.13 How interested would you say you personally are in politics?

PLEASE TICK ONE BOX ONLY

%

Very interested 8.1

Fairly interested 28.2

Somewhat interested 20.3

Not very interested 27.9

Not at all interested 15.1

Can't choose 0.4

NA 0.1

N15

2.14 Please tick one box on each line to show how much you agree or disagree with each of the following statements.

PLEASE TICK ONE BOX ON EACH LINE

m=620

	Strongly agree	Agree	Neither agree nor disagree	Disagree	Strongly disagree	Can't choose	NA
a. People like me don't have any say about what the government does	% 31.7	43.6	10.6	9.9	1.5	1.1	1.7
b. The average citizen has considerable influence on politics	% 3.2	10.3	16.7	48.3	16.8	1.8	3.0
c. Even the best politician cannot have much impact because of the way government works	% 9.9	46.3	21.2	14.7	1.2	4.1	2.7
d. I feel that I have a pretty good understanding of the important political issues facing our country	% 5.2	42.2	19.2	21.3	4.6	4.2	3.1
e. Elections are a good way of making governments pay attention to what the people think	% 14.7	46.0	20.2	13.3	2.4	1.7	1.6
f. I think most people are better informed about politics and government than I am	% 5.3	27.7	28.8	26.9	4.6	2.9	3.7
g. People we elect as MPs try to keep the promises they have made during the election	% 3.2	20.4	26.4	34.6	9.6	2.7	3.1
h. Most civil servants can be trusted to do what is best for the country	% 2.1	20.1	32.5	31.3	7.8	4.3	1.8

2.15 All in all, how well or badly do you think the system of democracy in the UK works these days?

PLEASE TICK ONE BOX ONLY

%

It works well and needs no changes 3.8

It works well but needs some changes 54.0

It does not work well and needs a lot of changes 22.8

It does not work well and needs to be completely changed 7.9

Can't choose 10.7

NA 0.7

And now some questions about taxes.

2.16 If the government had a choice between reducing taxes or spending more on social services which do you think it should do? (We mean all taxes together, including Income Tax, National Insurance, VAT and all the rest.)

PLEASE TICK ONE BOX ONLY

%

Reduce taxes, even if this means spending less on social services 21.4

OR

Spend more on social services, even if this means higher taxes? 54.8

Can't choose 22.8

NA 0.9

N16

2.17a Generally, how would you describe taxes in the UK today? First, for those with high incomes, are taxes ...

PLEASE TICK ONE BOX ONLY

n=620

	%
... much too high,	7.0
too high,	15.5
about right,	29.0
too low,	29.5
or, are they much too low?	10.5
Can't choose	6.3
NA	2.2

b. Next, for those with middle incomes, are taxes ...

PLEASE TICK ONE BOX ONLY

	%
... much too high,	8.7
too high,	30.2
about right,	45.5
too low,	7.6
or, are they much too low?	1.1
Can't choose	5.4
NA	1.5

c. Lastly, for those with low incomes, are taxes ...

PLEASE TICK ONE BOX ONLY

	%
... much too high,	35.1
too high,	39.7
about right,	17.7
too low,	0.5
or, are they much too low?	0.2
Can't choose	4.9
NA	1.8

2.18 Please tick one box on each line to show whether you think each of the following should mainly be run by private organisations or companies, or by government?

PLEASE TICK ONE BOX ON EACH LINE

	Mainly run by private organisations or companies	Mainly run by government	Can't choose	NA
Electricity	% 29.2	56.3	13.0	1.6
Hospitals	% 9.7	80.1	8.1	2.1
Banks	% 55.1	27.8	15.4	1.7

N17

2.19 Which of these statements comes closest to your view?

n=620

PLEASE TICK ONE BOX ONLY

	%
The UK's courts should be allowed to overrule parliament on any law which denies people their basic rights	56.5
OR	
The UK's democratically elected parliament should always have the final say on what the law should be	21.9
Can't choose	20.5
NA	1.2

2.20 How much do you agree or disagree with this statement?

"The UK should introduce proportional representation for Westminster elections, so that the number of MPs each party gets matches more closely the number of votes each party gets."

PLEASE TICK ONE BOX ONLY

	%
Strongly agree	21.9
Agree	36.4
Neither agree nor disagree	15.8
Disagree	6.0
Strongly disagree	2.0
Can't choose	17.1
NA	0.8

2.21 Here are some more ways people or organisations can protest against a government action they strongly oppose. Please show which you think should be allowed and which should not be allowed by ticking a box on each line.

PLEASE TICK ONE BOX ON EACH LINE

Should it be allowed?

	Definitely	Probably	Probably not	Definitely not	Can't choose	NA
a. Publishing pamphlets to protest against the government	% 33.3	41.3	10.1	6.0	7.6	1.7
b. Occupying a government office and stopping work there for several days	% 2.7	8.1	31.5	43.7	6.9	7.2
c. Seriously damaging government buildings	% 0.6	1.0	4.9	80.8	5.7	6.9

N18

n=620

2.22 Which of these statements comes closest to your view about general elections?

PLEASE TICK ONE BOX ONLY

%

In a general election ...

It's not really worth voting. 16.0

People should vote only if they care who wins 28.3

It is everyone's duty to vote 56.4

DK 0.1

NA 1.2

2.23 Here are some decisions that could be made either by the MPs we elect to parliament or by everyone having a say in a special vote or referendum.

a. First, who do you think should make the decision about whether or not the UK should replace the pound with a single European currency? Should the decision be made ...

%

PLEASE TICK ONE BOX ONLY

... by elected MPs in parliament 8.6

OR

by everyone in a referendum? 78.7

Can't choose 11.7

NA 1.1

b. And who do you think should make the decision about whether or not we should reintroduce the death penalty for some crimes? Should the decision be made ...

%

PLEASE TICK ONE BOX ONLY

... by elected MPs in parliament 15.5

OR

by everyone in a referendum? 71.4

Can't choose 11.9

NA 1.2

c. And who do you think should decide whether or not the UK should introduce proportional representation for Westminster elections so that the number of MPs each party gets matches more closely the number of votes each party gets? Should that decision be made ...

%

PLEASE TICK ONE BOX ONLY

... by elected MPs in parliament 11.6

OR

by everyone in a referendum? 67.9

Can't choose 19.4

NA 1.1

N19

n=620

2.24 Which of these statements about MPs comes closest to your view?

PLEASE TICK ONE BOX ONLY

%

It is a bad thing for MPs to have another paid job because being an MP is a full-time job in itself 68.6

OR

It is a good thing for MPs to have another paid job because it keeps them in touch with the outside world 20.4

Can't choose 10.2

NA 0.8

2.25 Please tick one box for each statement below to show how much you agree or disagree with it.

PLEASE TICK ONE BOX ON EACH LINE

	Strongly agree	Agree	Neither agree nor disagree	Disagree	Strongly disagree	Can't choose	NA
a. Any individual who gives money to a political party should be allowed to keep their gift private if they wish	% 10.1	36.5	15.7	17.6	10.0	6.3	3.8
b. There should be a limit on how much money a single individual can give to a political party	% 10.7	30.0	23.5	21.6	3.3	7.8	3.2
c. Political parties need to be funded by the government to do their job properly	% 7.7	26.2	27.7	21.9	4.8	9.7	2.0

N20

2.26 From what you know or have heard, please tick a box for each of the items below to show whether you think the National Health Service in your area is, on the whole, satisfactory or in need of improvement.

PLEASE TICK ONE BOX ON EACH LINE

n=620

	In need of a lot of improvement	In need of some improvement	Satisfactory	Very good	DK	NA
a. GPs' appointment systems	% 14.3	31.1	44.8	8.7	-	1.1
b. Amount of time GP gives to each patient	% 9.2	27.9	50.9	10.7	-	1.3
c. Being able to choose which GP to see	% 6.6	23.2	53.5	14.9	-	1.9
d. Quality of medical treatment by GPs	% 5.2	20.8	51.4	20.2	-	2.4
e. Hospital waiting lists for non-emergency operations	% 37.7	44.0	14.3	1.7	0.2	2.2
f. Waiting time before getting appointments with hospital consultants	% 44.3	39.2	12.7	2.0	-	1.8
g. General condition of hospital buildings	% 15.5	36.9	38.4	6.9	-	2.2
h. Staffing level of nurses in hospitals	% 40.0	33.4	20.3	4.4	-	1.8
i. Staffing level of doctors in hospitals	% 37.3	34.9	21.8	3.3	0.1	2.6
j. Quality of medical treatment in hospitals	% 11.3	24.8	46.5	15.2	-	2.2
k. Quality of nursing care in hospitals	% 9.3	18.0	42.4	27.9	-	2.3
l. Waiting areas in accident and emergency departments in hospitals	% 16.7	35.0	39.7	6.2	0.1	2.3
m. Waiting areas for out-patients in hospitals	% 15.9	29.6	44.7	7.2	0.1	2.5
n. Waiting areas at GPs' surgeries	% 4.0	18.1	62.3	13.7	-	2.0
o. Time spent waiting in out-patient departments	% 24.6	41.2	28.7	2.8	0.1	2.6
p. Time spent waiting in accident and emergency departments before being seen by a doctor	% 30.5	42.4	22.3	2.8	0.2	1.9
q. Time spent waiting for an ambulance after a 999 call	% 11.2	27.7	45.6	12.1	1.0	2.5

N21

2.27 In the last two years, have you or a close family member ...

PLEASE TICK ONE BOX ON EACH LINE

n=620

	Yes	No	NA
a. ... visited an NHS GP?	% 95.6	2.9	1.6
b. ... been an out-patient in an NHS hospital?	% 72.2	25.9	1.9
c. ... been an in-patient in an NHS hospital?	% 50.8	46.9	2.3
d. ... visited a patient in an NHS hospital?	% 83.9	13.1	3.0
e. ... had any medical treatment as a private patient?	% 9.0	88.7	2.3
f. ... had any dental treatment as a private patient?	% 18.2	79.6	2.3

2.28a Suppose two men with a heart condition go on a hospital waiting list at the same time. Both would benefit from an operation. One man does not smoke and the other smokes heavily. Who do you think would get the operation first.....

PLEASE TICK ONE BOX ONLY

%

...the non-smoker, 53.5
the heavy smoker, 2.7
or, would their smoking habits make no difference? 32.3
Can't choose 11.0
NA 0.6

b. And in your view, who do you think should get the operation first.....

PLEASE TICK ONE BOX ONLY

%

...the non-smoker, 37.9
the heavy smoker, 1.5
or, should their smoking habits make no difference? 51.1
Can't choose 8.8
NA 0.6

c. If decisions like this had to be made, who would you trust most to decide whether non-smokers or smokers should get the operation first?

PLEASE TICK ONE BOX ONLY

%

The government 2.9
Managers working for local health authorities 1.1
Managers in hospitals 3.6
Hospital doctors 77.1
Can't choose 14.5
NA 0.8

N22

n=820

2.29a Now suppose another two men with a heart condition go on a hospital waiting list at the same time. Both would benefit from an operation. One man is aged 40 and other aged 60. Who do you think would get the operation first....

PLEASE TICK ONE BOX ONLY

	%
...the younger man,	35.7
the older man,	6.7
or, would their ages make no difference?	46.1
Can't choose	10.4
NA	1.1

b And in your view, who do you think should get the operation first....

PLEASE TICK ONE BOX ONLY

	%
...the younger man,	20.1
the older man,	5.6
or, should their ages make no difference?	63.7
Can't choose	9.6
NA	1.1

2.30a Again suppose there are two men with a heart condition. One man is of average weight and eats healthily, whilst the other is very overweight and eats unhealthily. Again, both would benefit from an operation. Who do you think would get the operation first....

PLEASE TICK ONE BOX ONLY

	%
...the man whose weight is average and eats healthily,	44.8
the man who is very overweight and eats unhealthily,	2.6
or, would their weights make no difference?	40.8
Can't choose	10.7
NA	1.1

b. And in your view, who do you think should get the operation first....

PLEASE TICK ONE BOX ONLY

	%
...the man whose weight is average and eats healthily,	29.1
the man who is very overweight and eats unhealthily,	2.1
or, should their weights make no difference?	56.6
Can't choose	11.1
NA	1.1

N23

n=820

Now a few questions about the area where you live.

2.31 In some areas people do things together and try to help each other, while in other areas people mostly go their own way. In general, would you say you live in an area where ....

PLEASE TICK ONE BOX ONLY

	%
...people help each other,	40.0
OR people go their own way?	19.3
Mixture	39.3
Can't choose	0.5
NA	0.9

2.32a Do you think you live in the sort of area where people who thought a house was being broken into would ....

PLEASE TICK ONE BOX ONLY

	%
...do something about it,	66.8
OR just turn a blind eye?	1.9
Mixture	17.3
No burglaries in this area	11.5
Can't choose	1.6
NA	1.0

b And do you think burglaries in this area are ...

PLEASE TICK ONE BOX ONLY

	%
...mostly done by people from other areas,	40.2
OR mostly done by people from around here?	11.5
Mixture	18.3
No burglaries in this area	22.9
Can't choose	6.0
NA	1.1

2.33 Suppose a newly-married young couple, both with steady jobs, asked your advice about whether to buy or rent a home. If they had the choice, what would you advise them to do?

PLEASE TICK ONE BOX ONLY

	%
To buy a home as soon as possible	70.8
To wait a bit, then try to buy a home	22.7
Not to plan to buy a home at all	2.0
Can't choose	3.7
NA	0.8

N24

2.34 Still thinking of what you might say to this young couple, please tick one box for each statement below to show how much you agree or disagree with it.

n=620

PLEASE TICK ONE BOX ON EACH LINE

	Agree strongly	Just agree	Neither agree nor disagree	Just disagree	Disagree strongly	DK	NA
a. Owning your home can be a risky investment	% 7.8	27.2	19.0	27.2	16.5	0.1	2.2
b. Over time, buying a home works out less expensive than paying rent	% 54.0	34.7	5.2	2.8	1.2	0.1	2.0
c. Owning your home makes it easier to move when you want to	% 21.9	32.4	25.6	14.4	3.4	0.1	2.1
d. Owning a home ties up money you may need urgently for other things	% 5.8	23.5	31.2	27.4	9.2	0.1	2.8
e. Owning a home gives you the freedom to do what you want to it	% 34.8	36.7	17.0	6.3	2.7	-	2.5
f. Owning a home is a big financial burden to repair and maintain	% 13.4	35.2	24.9	17.3	5.9	0.1	3.2
g. Your own home will be something to leave your family	% 50.3	35.5	7.8	2.7	1.6	-	2.0
h. Owning a home is just too much of a responsibility	% 2.8	7.9	18.3	36.7	31.9	-	2.5
i. Owning a home is too much of a risk for couples without secure jobs	% 22.5	34.2	20.9	14.6	5.7	-	2.2
j. Couples who buy their own homes would be wise to wait before starting a family	% 13.7	26.2	36.6	14.4	7.0	-	2.2

2.35 Some people think that better relations between Protestants and Catholics in Northern Ireland will only come about through more mixing of the two communities. Others think that better relations will only come about through more separation. Which comes closest to your views ...

PLEASE TICK ONE BOX ONLY

%

Better relations will come about through more mixing   94.7
Better relations will come about through more separation   3.2
                                                    DK   0.3
                                                    NA   1.8

N25

2.36 And are you in favour of more mixing or more separation in ...

n=620

PLEASE TICK ONE BOX ON EACH LINE

	Much more mixing	Bit more mixing	Keep things as they are	Bit more separation	Much more separation	DK	NA
a. ... primary schools?	% 53.8	24.5	19.4	0.1	1.0	-	1.2
b. ... secondary and grammar schools?	% 50.7	30.6	16.0	0.1	1.2	-	1.4
c. ... where people live?	% 45.2	36.6	15.9	0.6	0.3	-	1.4
d. ... where people work?	% 52.7	33.7	11.5	0.4	0.1	-	1.6
e. ... people's leisure or sports activities?	% 54.9	30.4	13.5	-	0.1	-	1.1
f. ... people's marriages?	% 26.3	24.1	37.9	2.4	5.1	-	4.1

2.37 People feel closer to some groups than to others. For you personally, how close would you say you feel towards ...

PLEASE TICK ONE BOX ON EACH LINE

	Very close	Fairly close	A little close	Not very close	Not at all close	NA
a. ... people born in the same area as you?	% 18.3	46.0	19.5	11.4	3.0	1.8
b. ... people who have the same social class background as yours?	% 16.1	52.6	20.4	7.3	1.5	2.1
c. ... people who have the same religious background as yours?	% 17.2	47.0	22.2	8.8	2.8	2.0
d. ... people of the same race as you?	% 13.4	50.8	21.8	7.9	2.9	3.1
e. ... people who live in the same area as you do now?	% 13.6	50.1	25.1	7.2	2.6	1.3
f. ... people who have the same political beliefs as you?	% 9.8	39.8	28.3	13.5	5.8	2.8

N26

2.38 Please tick one box on each line to show how much you agree or disagree with each of the following statements.

PLEASE TICK ONE BOX ON EACH LINE

n=620

	Strongly agree	Agree	Neither agree nor disagree	Disagree	Strongly disagree	Can't choose	NA
a. Northern Ireland should remain part of the UK for as long as most of its people want it to do so	% 50.5	28.4	7.5	6.6	2.6	2.8	1.7
b. I would like the future of Northern Ireland to be within a United Ireland	% 9.9	12.7	23.4	18.7	28.3	4.8	2.2
c. Northern Ireland should become an independent country	% 4.1	8.9	22.4	30.1	24.9	6.8	2.8
d. Governing Northern Ireland should be done by Britain and Ireland together	% 8.0	16.5	18.8	27.0	21.8	5.5	2.2
e. It doesn't matter whether there is a United Ireland or whether Northern Ireland stays part of the UK as long as there is peace	% 29.9	21.5	14.0	14.3	13.5	5.1	1.7

2.39a How much say do you think a Westminster government of any party should have in the way Northern Ireland is run? Do you think it should have ...

PLEASE TICK ONE BOX ONLY

%
... a great deal of say, 17.6
some say, 43.3
a little say, 20.1
or - no say at all? 9.6
Can't choose 8.4
NA 1.0

b. And how much say do you think an Irish government of any party should have in the way Northern Ireland is run? Do you think it should have ...

PLEASE TICK ONE BOX ONLY

%
... a great deal of say, 6.8
some say, 21.3
a little say, 24.0
or - no say at all? 38.7
Can't choose 8.4
NA 0.8

N27

2.40 How proud are you of Northern Ireland in each of the following?

PLEASE TICK ONE BOX ON EACH LINE

n=620

	Very proud	Somewhat proud	Not very proud	Not proud at all	Can't choose	NA
a. Its economic achievements	% 21.4	50.6	14.2	2.6	8.9	2.4
b. Its achievements in sports	% 32.0	47.3	10.0	2.3	6.4	2.0
c. Its achievements in the arts and literature	% 23.5	42.4	13.8	2.8	15.5	2.0
d. Its fair and equal treatment of all groups in society	% 7.7	29.1	33.0	17.6	10.4	2.2

Now some questions about the countryside.

2.41a Which one of these two statements comes closest to your own views?

PLEASE TICK ONE BOX ONLY

%

Industry should be prevented from causing damage to the countryside, even if this sometimes leads to higher prices 91.0

OR

Industry should keep prices down, even if this sometimes causes damage to the countryside 6.2

DK 0.4
NA 2.4

b. And which of these two statements comes closest to your own views?

PLEASE TICK ONE BOX ONLY

%

The countryside should be protected from development, even if this sometimes leads to fewer jobs 71.7

OR

New jobs should be created, even if this sometimes causes damage to the countryside 25.6

DK 0.5
NA 2.3

N28

2.42 Please tick one box on each line to show how you feel about....

PLEASE TICK ONE BOX ON EACH LINE

n=620

	It should be stopped altogether	It should be discouraged	Don't mind one way or the other	It should be encouraged	DK	NA
a. ... increasing the amount of countryside being farmed	% 3.2	23.5	47.7	22.4	0.1	3.1
b. ... building new housing in country areas	% 9.8	41.7	33.8	12.2	0.1	2.3
c. ... putting the needs of farmers before protection of wildlife	% 13.8	49.9	25.4	8.1	0.1	2.6
d. ... providing more roads in country areas	% 10.0	35.7	33.1	18.0	0.1	3.1
e. ... increasing the number of picnic areas and camping sites in the countryside	% 3.0	13.1	34.4	47.1	0.1	2.2

2.43a The new owner of a stately home containing historic paintings and furniture wishes to close it to the public. Should he or should he not have the right to do this?

PLEASE TICK ONE BOX ONLY

%
Definitely should have the right 29.1
Probably should have the right 34.1
Probably should not have the right 9.8
Definitely should not have the right 5.7
It depends 14.3
Can't choose 5.7
NA 1.2

2.43b A new landowner of a large estate in a beautiful part of the UK decides to fence off a remote part of his land to stop people visiting it. Should he or should he not have the right to do this?

PLEASE TICK ONE BOX ONLY

%
Definitely should have the right 25.7
Probably should have the right 38.5
Probably should not have the right 12.6
Definitely should not have the right 7.1
It depends 10.8
Can't choose 4.0
NA 1.3

N29

2.43c Suppose a rarely-used public footpath runs through farming land. Should the farmer be able to get it closed without a lot of fuss and bother?

PLEASE TICK ONE BOX ONLY

n=620

%
Definitely should have the right 16.1
Probably should have the right 26.8
Probably should not have the right 24.0
Definitely should not have the right 16.2
It depends 12.9
Can't choose 3.5
NA 0.6

And now some questions about the environment.

2.44 How much do you agree or disagree with each of these statements?

PLEASE TICK ONE BOX ON EACH LINE

	Strongly agree	Agree	Neither agree nor disagree	Disagree	Strongly disagree	Can't choose	NA
a. It is just too difficult for someone like me to do much about the environment	% 8.8	34.3	21.3	25.2	5.1	2.7	2.6
b. I do what is right for the environment, even when it costs more money or takes more time	% 7.9	44.5	28.3	7.8	0.9	5.5	5.1

2.45a In general, do you think that air pollution caused by cars is ...

PLEASE TICK ONE BOX ONLY

%
... extremely dangerous for the environment, 22.0
very dangerous, 28.1
somewhat dangerous, 40.0
not very dangerous, 6.3
or, not dangerous at all for the environment? 0.8
Can't choose 2.2
NA 0.6

N30

2.45b And do you think that air pollution caused by cars is ....

PLEASE TICK ONE BOX ONLY

n=820

%

...extremely dangerous for you and your family?	20.0
very dangerous,	23.8
somewhat dangerous,	43.5
not very dangerous,	8.6
or, not dangerous at all for you and your family?	0.7
Can't choose	2.8
NA	0.6

c. Within the next ten years, how likely do you think it is that there will be a large increase in ill-health in the UK's cities as a result of air pollution caused by cars?

PLEASE TICK ONE BOX ONLY

%

Certain to happen	26.4
Very likely to happen	24.9
Fairly likely to happen	30.5
Not very likely to happen	11.8
Certain not to happen	0.2
Can't choose	5.5
NA	0.7

2.46 Please tick the box that comes closest to your opinion of how true this statement is.

"Cars are not really an important cause of air pollution the UK."

PLEASE TICK ONE BOX ONLY

%

Definitely true	5.8
Probably true	24.7
Probably not true	28.0
Definitely not true	36.0
Can't choose	5.0
NA	0.6

N31

2.47 If you had to choose, which one of the following would be closest to your views?

n=820

%

PLEASE TICK ONE BOX ONLY

Government should let ordinary people decide for themselves how to protect the environment, even if it means they don't always do the right thing	22.2
OR	
Government should pass laws to make ordinary people protect the environment, even if it interferes with people's rights to make their own decisions	51.6
Can't choose	23.7
NA	2.4

2.48 And which one of the following would be closest to your views?

%

PLEASE TICK ONE BOX ONLY

Government should let businesses decide for themselves how to protect the environment, even if it means they don't always do the right thing	6.8
OR	
Government should pass laws to make businesses protect the environment, even if it interferes with business' rights to make their own decisions	75.5
Can't choose	15.2
NA	2.6

2.49 On the whole, which of these statements comes closest to your own views?

%

PLEASE TICK ONE BOX ONLY

It's mainly up to the government to protect the environment - ordinary people can't do much on their own	48.5
OR	
It's mainly up to ordinary people to do what they can to protect the environment - the government can do only a limited amount	48.1
Can't choose	0.4
NA	3.0

2.50 Please tick one box for each statement below to show how much you agree or disagree with it.

PLEASE TICK ONE BOX ON EACH LINE

		Agree strongly	Agree	Neither agree nor disagree	Disagree	Disagree strongly	DK	NA
a.	The government should do more to protect the environment, even if it leads to higher taxes	% 8.8	46.6	31.7	8.0	1.2	0.2	3.5
b.	Industry should do more to protect the environment, even if it leads to lower profits and fewer jobs	% 11.4	55.9	23.1	4.6	0.9	0.2	3.8
c.	Ordinary people should do more to protect the environment, even if it means paying higher prices	% 9.3	49.5	26.7	10.1	0.8	0.2	3.4
d.	People should be allowed to use their cars as much as they like, even if it causes damage to the environment	% 1.5	15.6	35.3	37.8	6.1	0.2	3.5

N32

Now, two questions on roads and public transport.

2.51a Thinking first about towns and cities. If the government had to choose ...

n=620

%

PLEASE TICK ONE BOX ONLY

It should improve roads	33.8
It should improve public transport	63.2
NA	3.0

b. And in country areas, if the government had to choose ...

%

PLEASE TICK ONE BOX ONLY

It should improve roads	42.9
It should improve public transport	53.6
NA	3.5

2.52 How much trust do you have in each of the following groups to help the UK make the right decisions about the environment?

PLEASE TICK ONE BOX ON EACH LINE

	A lot of trust	Some trust	Very little trust	No trust at all	Can't choose	NA
a. Scientists	15.4	53.7	14.4	5.5	6.3	4.8
b. Business and industry	1.2	27.2	46.1	16.5	4.4	4.6
c. Environmental groups	37.4	46.1	6.3	2.0	4.4	3.9
d. The government	3.0	48.3	31.8	9.6	4.2	3.1
e. Ordinary people	10.4	52.8	25.1	2.4	5.4	4.0

N33

2.53 Please tick one box for each statement to show how much you agree or disagree with it.

n=620

PLEASE TICK ONE BOX ON EACH LINE

	Agree strongly	Agree	Neither agree nor disagree	Disagree	Disagree strongly	NA
a. The welfare state makes people nowadays less willing to look after themselves	% 11.0	31.7	23.0	28.5	4.2	1.6
b. People receiving social security are made to feel like second class citizens	% 11.7	39.6	23.4	21.4	2.6	1.2
c. The welfare state encourages people to stop helping each other	% 3.7	27.5	32.2	32.2	2.4	1.9
d. The government should spend more money on welfare benefits for the poor, even if it leads to higher taxes	% 14.1	40.4	26.8	14.9	2.8	1.0
e. Around here, most unemployed people could find a job if they really wanted one	% 10.4	33.2	22.6	27.3	5.6	1.0
f. Many people who get social security don't really deserve any help	% 3.7	24.9	24.1	37.8	8.2	1.3
g. Most people on the dole are fiddling in one way or another	% 11.5	29.0	28.7	23.4	6.0	1.3
h. If welfare benefits weren't so generous, people would learn to stand on their own two feet	% 7.2	24.0	24.5	32.2	10.6	1.5

2.54 Please tick one box for each statement below to show how much you agree or disagree with it.

PLEASE TICK ONE BOX ON EACH LINE

	Agree strongly	Agree	Neither agree nor disagree	Disagree	Disagree strongly	Can't choose	NA
a. Ordinary working people get their fair share of the nation's wealth	% 2.2	11.9	26.0	47.3	11.5	-	1.1
b. There is one law for the rich and one for the poor	% 24.9	47.9	15.6	9.7	1.0	-	0.9
c. Young people today don't have enough respect for traditional values	% 22.9	51.7	15.3	7.4	1.6	-	1.1
d. Censorship of films and magazines is necessary to uphold moral standards	% 25.2	45.1	14.7	9.6	3.6	-	1.7
e. There is no need for strong trade unions to protect employees' working conditions and wages	% 2.8	14.4	23.6	45.4	12.6	-	1.2
f. Private enterprise is the best way to solve the UK's economic problems	% 4.3	20.4	45.7	22.3	5.0	-	2.3
g. Major public services and industries ought to be in state ownership	% 11.9	28.2	40.8	15.5	1.9	0.1	1.7

N34

2.55 Please tick one box for each statement below to show how much you agree or disagree with it.

PLEASE TICK ONE BOX ON EACH LINE

n=620

	Agree strongly	Agree	Neither agree nor disagree	Disagree	Disagree strongly	NA
a. It is the government's responsibility to provide a job for everyone who wants one	% 18.8	39.2	22.2	16.1	1.5	2.2
b. People should be allowed to organise public meetings to protest against the government	% 11.0	53.7	22.1	8.6	1.7	2.9
c. Homosexual relations are always wrong	% 23.5	19.1	28.0	20.5	6.1	2.8
d. People in the UK should be more tolerant of those who lead unconventional lives	% 8.2	39.7	36.1	9.4	3.8	2.8
e. Political parties which wish to overthrow democracy should be allowed to stand in general elections	% 2.6	15.6	29.5	34.2	15.3	2.7

2.56 Please tick one box for each statement below to show how much you agree or disagree with it.

PLEASE TICK ONE BOX ON EACH LINE

	Agree strongly	Agree	Neither agree nor disagree	Disagree	Disagree strongly	NA
a. Government should redistribute income from the better-off to those who are less well off	% 15.1	38.3	25.1	17.7	1.8	1.8
b. Big business benefits owners at the expense of workers	% 15.8	48.9	23.3	9.1	0.9	2.0
c. Ordinary working people do not get their fair share of the nation's wealth	% 18.3	54.2	17.6	7.2	1.0	1.6
d. Management will always try to get the better of employees if it gets the chance	% 17.6	51.6	20.1	8.4	0.7	1.7
e. People who break the law should be given stiffer sentences	% 30.4	47.3	17.5	2.6	0.6	1.6
f. For some crimes, the death penalty is the most appropriate sentence	% 28.2	25.1	18.8	14.9	11.1	1.9
g. Schools should teach children to obey authority	% 35.7	51.5	10.1	1.2	-	1.5
h. The law should always be obeyed, even if a particular law is wrong	% 8.8	28.3	30.9	25.8	4.5	1.7

N35

n=620

2.57a To help us plan better in future, please tell us about how long it took you to complete this questionnaire.

PLEASE TICK ONE BOX ONLY

	%
Less than 15 minutes	3.9
Between 15 and 20 minutes	21.0
Between 21 and 50 minutes	33.0
Between 31 and 45 minutes	20.1
Between 46 and 60 minutes	13.1
Over one hour	7.9
NA	1.0

b. And on what date did you fill in the questionnaire?

PLEASE WRITE IN: DATE   MONTH   1996

Thank you very much for your help

Please keep the completed questionnaire for the interviewer if he or she has arranged to call for it. Otherwise, please post it as soon as possible in the pre-paid envelope provided.

Printed and bound by CPI Group (UK) Ltd, Croydon, CR0 4YY

21/10/2024

01777087-0004